STATISTICS   Made Simple

H. T. Hayslett, MS

Advisory editor
Patrick Murphy, MSc, FIMA

Made Simple Books
W. H. ALLEN   London
A Howard & Wyndham Company

© 1968 edition by W. H. Allen & Co., Ltd.

Made and printed in Great Britain
by Richard Clay (The Chaucer Press), Ltd, Bungay, Suffolk
for the publishers W. H. Allen & Co., Ltd.
44 Hill Street, London W1X 8LB

First edition, April 1968
Reprinted, August 1970
Second (revised) edition, May 1971
Reprinted, April 1973
Third (revised) edition, November 1974
Reprinted, September 1976
Reprinted, January 1978
Reprinted, May 1979

ISBN 0 491 01811 8 Paperbound

STATISTICS Made Simple

The Made Simple series
has been created
primarily for self-education
but can equally well
be used as
an aid to group study.
However complex the subject,
the reader is taken
step by step,
clearly and methodically
through the course. Each volume
has been prepared by
experts,
using throughout the
Made Simple technique of teaching.
Consequently the gaining
of knowledge now becomes
an experience to be enjoyed.

Foreword

The use of statistical methods is becoming increasingly important in all fields of everyday enterprise. Indeed, it is almost impossible to name an activity which does not employ its own particular Statistics as an aid to influencing human behaviour.

But the usefulness of any statistical enquiry or market research depends entirely on the competence of those who attempt to interpret the results they obtain. Since no commercial undertaking of the present day is prepared to tolerate the doubtful judgements of mere amateurs applying 'rules of thumb', it follows that there is a greater than ever demand for those who are better informed about the techniques and methods of Statistics. A great deal of responsibility rests upon the shoulders of anyone acting in an advisory capacity related to the findings of a statistical enquiry, a responsibility which this book will enable him to assume with authority and confidence.

In this book Professor Hayslett has presupposed no mathematical training, except school algebra, on the part of his readers. Possibly summation notation, inequalities and the equations of straight lines will be unfamiliar to some students. If so, they will find the topics treated where they are needed, making the book quite self-contained.

Numerous examples are included but since Statistics is such an immense field it has proved impossible to draw examples from all subject areas. The author has assumed that any one using this book will do so because he must have some knowledge of the subject for his work or study. The reader can, therefore, undoubtedly supply his own specific applications after reading the explanation and examples given. In later chapters, where the problems become more complex, step-by-step directions for the various statistical tests are given and these directions are illustrated by examples. Few books carry such a detailed study of these tests.

The material in Chapters 2, 3 and 4 is basic, and is needed throughout the book. The material in Chapter 5 is needed specifically for Chapter 6. More generally, a knowledge of probability is helpful throughout (see *Mathematics Made Simple* in the same series), but it is *not essential* for enabling the student to perform the various tests.

This book contains a clear presentation of the basic principles of Statistics which any reader, regardless of his own personal speciality, will find very helpful. It can therefore be fully recommended as a companion volume for any student taking one of the academic courses leading to G.C.E. at both O and A levels.

PATRICK MURPHY

Table of Contents

CHAPTER I

WHAT IS STATISTICS?

In order to study the subject of statistics intelligently we should first understand what the term means today, and know something of its origin.

As with most other words, the word 'statistics' has different meanings for different people. When most people hear the word they think of tables of figures giving births, deaths, marriages, divorces, car accidents, and so on. This is indeed a vital and correct use of the term. In fact, the word 'statistics' was first applied to these affairs of the State, to data that government finds necessary for effective planning, ruling, and tax-collecting. Collectors and analysers of this information were once called 'statists', which shows much more clearly than the term 'statistician' the original preoccupation with the facts of the state.

Today, of course, the term 'statistics' is applied, in this first sense, to nearly any kind of factual information given in terms of numbers—the so-called 'facts and figures'. Radio and television announcers tell us that they will 'have the statistics of the game in a few minutes', and newspapers frequently publish articles about beauty contests giving the 'statistics' of the contestants.

The term 'statistics', however, has other meanings, and people who have not studied the subject are relatively unfamiliar with these other meanings. Statistics is a body of knowledge in the area of applied mathematics, with its own symbolism, terminology, content, theorems, and techniques. When people study the subject, statistics, they usually attempt to master some of these techniques.

The *term* 'statistics' has a second meaning for those who have been initiated into the mysteries of the *subject* 'statistics'. In this second sense, 'statistics' are quantities that have been calculated from sample data; a single quantity that has been so calculated is called a 'statistic'. For example, the sample mean is a statistic, as are the sample median and sample mode (all discussed in Chapter III). The sample variance is a statistic, and so is the sample range (both discussed in Chapter IV). The sample correlation coefficient (discussed in Chapter IX) is a statistic, and so on.

We can summarize these meanings of the word 'statistics':

1. The public meaning of facts and figures, graphs, and charts. The word is plural when used in this sense.
2. The subject itself, with a terminology, methodology, and body of knowledge of its own. The word is singular when used in this sense.
3. Quantities calculated from sample data. The word is plural when used in this sense.

In this book we will not use the word 'statistics' at all in the first sense above. When we want to refer to 'facts and figures' we will use the term 'observations', or the term 'data'. We will occasionally refer to a quantity that has been calculated from sample data as a 'statistic'. In these cases we will be using the singular of the word 'statistics', in the third sense above. *Nearly always, when we use the word 'statistics' we will mean the subject itself, the body of knowledge.*

The methodology of statistics is sufficiently misunderstood to give rise to a number of humorous comments about statistics and statisticians. For example: 'A statistician is a person who draws a mathematically precise line from an unwarranted assumption to a foregone conclusion.' This strikes out at two abuses of statistical techniques, although the abuse is not by professional statisticians. In order to apply most statistical techniques, certain assumptions must be made, the number and scope of the assumptions varying from situation to situation. Perhaps some persons do make assumptions that they know are not justified, and disguise their doubt. And perhaps, also, some persons *do* have a conclusion already decided upon, and then choose their sample or 'doctor' their data in order to 'prove' their conclusion. Each of these abuses, when knowingly done, is dishonest.

One indictment of the techniques and methodology of statistics says that 'statistical analysis has often meant the manipulation of ambiguous data by means of dubious methods to solve a problem that has not been defined'.

Probably the remark that is known best is the one attributed by Mark Twain to Disraeli: 'There are three kinds of lies: lies, damned lies, and statistics.' And yet another well-known remark, critical of the manner in which statistics are used, is, 'He uses statistics as a drunk uses a street lamp—for support, rather than illumination.'

THE PRESENT IMPORTANCE OF STATISTICS

The application of statistical techniques is so widespread, and the influence of statistics on our lives and habits so great, that the importance of statistics can hardly be over-emphasized.

Our present agricultural abundance can be partially ascribed to the application of statistics to the design and analysis of agricultural experiments. This is an area in which statistical techniques were used relatively early. Some questions that the methods of statistics help to answer are: Which type of corn gives the best yield? Which feed mixture should chickens be fed so that they will gain the most weight? What kind of mixture of grass seeds gives the most tons of hay per are? All of these questions, and hundreds of others, have a direct effect on all of us through the local supermarket.

The methodology of statistics is also used constantly in medical and pharmaceutical research. The effectiveness of new drugs is determined by experiments, first on animals, and then on humans. New developments in medical research and new drugs affect most of us.

Statistics is used by the Government as well. Economic data are studied and affect the policies of the government in the areas of taxation, funds spent for public works (such as roads, bridges, etc.), public assistance funds, and so on. Statistics on unemployment affect efforts to lower the unemployment rate. Statistical methods are used to evaluate the performance of every sort of military equipment, from bullets used in pistols to huge missiles. Probability theory and statistics (especially a rather new area known as statistical decision theory) are used as an aid in making extremely important decisions at the highest levels.

In private industry the uses of statistics are nearly as important and their effects nearly as widespread as they are in government use. Statistical techniques are used to control the quality of products being produced and to evaluate new products before they are marketed. Statistics are used in marketing, in decisions to expand business, in the analysis of the effectiveness of advertising, and so on. Insurance companies make use of statistics in establishing their rates at a realistic level.

The list could go on and on. Statistics is used in geology, biology, psychology, sociology—in any area in which decisions must be made on the basis of incomplete information. Statistics is used in educational testing, in safety engineering. Meteorology, the science of weather prediction, is using statistics now.

On the lighter side, statistical studies have been made of the effect of the full moon on trout fishing; of which of two kinds of water glasses are better for use in restaurants; and of the optimum strategies for games of skill and chance such as bridge.

There can be little doubt, then, of the effect of statistics and statistical techniques on each of us. The results of statistical studies are seen, but perhaps not realized, in our wage packets, our national security, our insurance premiums, our satisfaction with products of many kinds, and our health.

TWO KINDS OF STATISTICS

In addition to a brief consideration of the basic elements of probability, there are two kinds of statistics treated in this book. In Chapters II, III, and IV we are concerned primarily with the description of data. In Chapter II we treat the pictorial description of data; in Chapters III and IV we treat the numerical description of data. The natural name for this kind of statistics is **descriptive statistics**. The classification of data; the drawing of histograms that correspond to the frequency distributions that result after the data are classified; the representation of data by other sorts of graphs, such as line graphs, bar graphs, pictograms; the computation of sample means, medians, or modes; the computation of variances, mean absolute deviations, and ranges—all these activities deal with descriptive statistics. The statistical work done back in the nineteenth century and the early part of this century was largely descriptive statistics.

The second important kind of statistics is known as **inferential statistics**. Statistics has been described as the science of making decisions in the face of uncertainty; that is, making the best decision on the basis of incomplete information. In order to make a decision about a population, a sample (usually just a few members) of that population is selected from it. The selection is usually by a random process. Although there are various kinds of sampling, the kind that we will be assuming throughout this book is known as random sampling. As the term suggests, this is a kind of sampling in which the members of the sample are selected by some sort of process that is not under the control of the experimenter. There are various mathematical definitions of random sampling, but we will consider it as a sample for which each member of the population has an equal chance of being selected, and for which the selection of any one member does not affect the selection of any other member.

On the basis of the random sample, we *infer* things about the population. This inferring about populations on the basis of samples is known as **statistical inference**. In other words, statistical inference is the use of samples to reach conclusions about the populations from which those samples have been drawn.

Let us mention several examples of statistical inference. Suppose that a manufacturer of tricycles buys bolts in large quantities. The manufacturer has the right to refuse to accept the consignment if more than 3 per cent of the bolts are defective. It is not feasible, of course, to check all of the bolts before they are used. This would take too long. Neither is it possible to simply lay aside the defective bolts as they are encountered during the assembly of the tricycles. The bolts cannot be returned after they have been used, even if 20 per cent are defective; and, of course, the tricycle manufacturer does not want to use a consignment of bolts that contains a large percentage of defectives, because it is expensive to attempt to use a defective bolt, realize that it is defective, and then do the job again with a satisfactory bolt. So for several reasons, the manufacturer needs to have a quick, inexpensive method by which he can determine whether the consignment contains too many defectives. So he obtains a random sample from the consignment of bolts, and on the basis of the percentage of defectives in the sample, he makes a decision about the percentage of defectives in the population (the consignment). This is an example of statistical inference.

Consider another example of statistical inference. A medical research worker wants to determine whether a new drug is superior to the old one. One hundred patients in a large hospital are divided at random into two groups. One group is given the old drug and the other group is given the new drug. Various medical data are obtained for each patient on the day the administration of the drug began, and the same things measured ten days later. By analysing the data for each group, and by comparing the data, a conclusion may be reached about the relative effectiveness of the two drugs.

A similar example is discussed at greater length at the beginning of Chapter VII, in which the testing of statistical hypotheses is first discussed.

The usual procedure for testing a statistical hypothesis is the following: A hypothesis, known as the null hypothesis, is proposed about a population; a random sample is obtained from the population, and a numerical quantity, known as a statistic, is calculated from the sample data. The null hypothesis is accepted or rejected, depending upon the value of the statistic. (An alternative hypothesis is formulated at the time as the null hypothesis, and rejection of the null hypothesis means automatic acceptance of the alternative hypothesis.) Thus, the testing of a statistical hypothesis is an illustration of statistical inference, because a decision is made about a population by means of a sample. Chapters VII, VIII, IX, XI, and XII are concerned (some partially, some entirely) with tests of statistical hypotheses, and therefore with statistical inference. Chapter X, in which an important topic known as confidence intervals is discussed, also deals with statistical inference.

In summary, the subject matter in this book falls rather naturally into three categories. Chapters II, III, and IV treat the description of data, both graphically and numerically, and are classified as descriptive statistics. Some very simple topics from probability theory are discussed in Chapters V and VI:

elementary probability, two important distributions—the binomial and the normal—and how the two are related, and the use of the normal table. The final six chapters treat selected topics, mainly about testing hypotheses, from that part of the subject matter of statistics known as inferential statistics. The first two categories, comprising the first six chapters, are preliminary to the last one.

CHAPTER II

PICTORIAL DESCRIPTION OF DATA

INTRODUCTION

This chapter is concerned with the presentation of sample data. Before treating the classification of data and the sketching of histograms we will briefly discuss the idea of a random sample and how one can be obtained.

If one is sampling from a population composed of an infinite number of elements, a sample selected in such a manner that the selection of any member of the population does not affect the selection of any other member, and each member has the same chance of being included in the sample, is called a **random sample**. If one is sampling from a finite population with replacement (each member is returned to the population after being selected, and might be selected more than once), a random sample is defined exactly as above. In other words, a finite population in which the sampling is carried out with replacement can, theoretically, be considered as an infinite population.

If one is sampling from a finite population without replacement (the elements are not returned to the population after they have been observed), then we say that it is a random sample if all other samples of the same size have an equal chance of being selected. No sample is any more likely to be selected than any other. For example, a choice of three raffle tickets from a bag containing 100 tickets is a sample from a finite population without replacement. It is a random sample because any other three tickets had the same chance of being selected.

The word 'random' indicates that the sample is selected in such a way that it is impossible to predict which members of the population will be included, and that it is simply a matter of chance that any particular member is selected. In order to apply the statistical techniques explained in this book in analysing sample data, it is necessary that the sample be a random one (with very few exceptions). The statistical techniques are justified by statistical theory, which in turn rests upon probability theory, and we must have random samples before the probability theory is applicable.

SELECTING A RANDOM SAMPLE

It is sometimes not an easy matter to obtain a random sample. If the population is small, one of the simplest ways of obtaining a random sample is to list the members (on small pieces of paper, for instance) and draw the sample 'out of a hat'.

Whenever an integer can be assigned to each member of the population, a **random-number table** can be used to obtain a random sample. This table is a listing of digits that have been obtained by some random process. One way of assigning an integer to each member of the population is simply to number the members 1, 2, 3, and so on. (Sometimes the members cannot be conveniently numbered, in which case there are other methods of obtaining a random sample by means of random numbers.) Each member of the population has a corresponding number in the random-number table (or perhaps more than one corresponding number). To obtain a random sample, we would begin reading numbers in the random-number table at some randomly chosen place, and for each random number read, the member of the population that corresponds to that number is included in the sample. For instance, if our population consists of a thousand members we could assign them numbers from 000 to 999. If we read the numbers 027, 831, and 415 in the random-number table we would include in the random sample those members of the population whose numbers are 027, 831, and 415.

The data shown in Table 2.1 are the scores that one hundred students obtained on the verbal portion of a Scholastic Aptitude Test; we shall refer to these scores as the SAT-Verbal scores. The sample was obtained from a population of students, using a table of random numbers to guarantee that the sample was random.

Table 2.1
Random Sample of 100 SAT-Verbal Scores

546	592	591	602	619
689	644	546	602	695
490	536	618	669	599
531	586	622	689	560
603	555	464	599	618
549	612	641	597	622
663	546	534	740	644
515	496	503	599	618
557	631	502	605	547
673	708	624	528	645
650	656	599	586	536
546	515	644	599	734
502	541	530	663	599
547	579	666	578	635
496	541	605	560	695
426	555	483	641	546
515	609	534	645	572
637	457	631	721	578
541	592	666	619	663
547	624	567	489	528

CLASSIFICATION OF DATA

As we examine these data, it is difficult to tell, without lengthy scrutiny, just how they are distributed. We find, after some searching, that the smallest observation is 426 and the largest observation is 740; also, it becomes apparent that there are few observations below 500 or above 700. But we cannot quickly tell whether there are as many observations between 500 and 550 as between 650 and 700. We need to arrange the data so that the main features will be clear. When data are arranged in order from smallest to largest we have what is known as an **array**. The array for the data in Table 2.1 is given in Table 2.2.

Now, it is obvious, after a brief examination of the array, that the observations in the 500s make up about half the 100 observations, that the observations in the 600s account for about another 40 per cent, and that observations less than 500 or greater than 700 account for only about 10 per cent. We are able to learn more with less effort than we were when the data were not arranged.

Table 2.2
SAT-Verbal Data—Arranged in Order

426	536	572	605	644
457	536	578	609	645
464	541	578	612	645
483	541	579	618	650
489	541	586	618	656
490	546	586	618	663
496	546	591	619	663
496	546	592	619	663
502	546	592	622	666
502	546	597	622	666
503	547	599	624	669
515	547	599	624	673
515	547	599	631	689
515	549	599	631	689
528	555	599	635	695
528	555	599	637	695
530	557	602	641	708
531	560	602	641	721
534	560	603	644	734
534	567	605	644	740

But still the data must be studied in order to draw these conclusions. Many persons do not like to examine a mass of numbers, and many others don't have the time to do so. Therefore, it would be advantageous if the information present in the array of observations could somehow be 'compressed' so that the distribution of the observations could be seen at a glance.

The device of **classifying the data** is used to 'compress' the data. The range of the observations (in this case $740 - 426 = 314$) is divided into a number of **class intervals**, or simply **classes**. Although the class intervals do not have to be equal, there are important advantages if they are; consequently, we will use equal class intervals exclusively in this book.

We must decide how many classes we wish to have. For large samples (over fifty observations, say) from ten to twenty classes will usually do nicely. For smaller samples fewer classes can be used—as few as five or six, perhaps. It should be emphasized that the number of classes is arbitrary. Given the same data, one person might classify them into twelve classes, another into fourteen, and yet another into only nine. In most problems (assuming a large number of observations) fewer than ten classes will result in too much information being lost; and if more than twenty are used the work involved in analysing the data becomes more and more lengthy.

But let us return to the problem of deciding what the value of k, the number of classes, should be here. The range is 314 units. If we use ten classes the width of each class interval would be 31·4 units; if twenty classes are used the width of each class interval would be 15·7 units. Any convenient number between 15·7 and 31·4 will do for the width of the class interval. We will use 13 classes, each of width 25 units.

Just as the number of classes and the width of the class intervals are arbitrary, so also is the point at which to begin the lowest class. We could begin the first class at 425. Thus the first class would be from 425 to 450, the second from 450 to 475, the third from 475 to 500, and so on. The numbers 425, 450, 475, and 500 are known as **class boundaries**: they separate one class from another. These boundaries are not well chosen, however, because it is not clear what should be done with certain values, such as 475. Should we put 475 into the lower class, into the upper class, or into both? The difficulty is not serious, and can be avoided if we specify the classes like this: 425 but not 450, 450 but not 475, 475 but not 500, and so on.

Another way around this difficulty is to use class boundaries which are more accurate than the observations. If the observations are given to the nearest integer the boundaries should be given correct to the nearest half; if the observations are given correct to tenths, then the boundaries should be given correct to twentieths, and so on. Using this procedure, boundaries for the first three classes would be (arbitrarily beginning at 424·5) 424·5–449·5, 449·5–474·5, 474·5–499·5.

The smallest and largest possible measurements in each class are called the **class limits**. Classes are sometimes specified in terms of the class limits. If this is done there is no overlap as there was in the first example of selecting class boundaries, because the largest possible observation in one class cannot be the smallest possible in another class. Specified in terms of their class limits, the first three classes would be 425–449, 450–474, and 475–499. If the scores had been reported to the nearest tenth of a unit, then scores of 449·9 as well as 425·0 would be possible. With this more accurate measurement the class limits of the first three classes would be 425·0–449·9, 450·0–474·9, and 475·0–499·9.

When the classes are described in terms of the class limits each boundary is understood to be half-way between the upper class limit of the lower class and the lower class limit of the upper class. For the class limits 425–449, 450–474, and 475–499 the class boundaries are 424·5, 449·5, 474·5, and 499·5.

The midpoint of a particular class interval is the point half-way between the class boundaries of that class, and is called the **class mark**. If the class boundaries are 424·5–449·5, 449·5–474·5, 474·5–499·5, . . . then the class marks are 437, 462, 487, . . . If the class boundaries are 424·95–449·95, 449·95–474·95, 474·95–499·95, . . . then the class marks would be 437·45, 462·45, 487·45, . . .

NOTE: Later, when we calculate the mean and the variance for classified data, we will see that the class mark is important—each observation in a particular class is assumed to have the same value as the class mark.

Summarizing these results we have:

    (i) Class interval   425–449
    (ii) Class limits   lower limit 425, upper limit 449
    (iii) Class boundary 424·5, 449·5

    (iv) Class mark $\dfrac{424·5 + 449·5}{2} = 437$

The number of observations in any particular class is called the **class frequency** of that class. The class frequency of the ith class (there are k classes,

Statistics Made Simple

so i can be any integer from 1 to k) is denoted by f_i. Thus f_1 is the class frequency of the first class, f_2 that of the second class, and so on. Since there are k classes, class frequency of the last class is denoted f_k.

FREQUENCY DISTRIBUTIONS AND CUMULATIVE FREQUENCY DISTRIBUTIONS

When the class intervals (in terms of class boundaries, limits, or marks) are displayed in tabular form along with the corresponding class frequencies, the resulting table is known as a **frequency distribution**.

The classes in the table below are given in terms of their class limits. Since we have already arranged the data in order, counting the number in each class can be done very quickly. The frequency distribution for the SAT-Verbal data is shown in Table 2.3.

Table 2.3

Class	Class limits	Class frequency
1	425–449	1
2	450–474	2
3	475–499	5
4	500–524	6
5	525–549	20
6	550–574	7
7	575–599	15
8	600–624	16
9	625–649	11
10	650–674	9
11	675–699	4
12	700–724	2
13	725–749	2

If the data are not arranged in order, then probably the quickest way to classify them is to read down the list of the unordered data, placing a tally mark opposite the appropriate class for each observation. Then the tallies are

Table 2.4

Class	Class limits	Tallies	Class frequency
1	425–449			1
2	450–474				2
3	475–499	⊥⊥⊥⊥	5
4	500–524	⊥⊥⊥⊥		6
5	525–549	⊥⊥⊥⊥ ⊥⊥⊥⊥ ⊥⊥⊥⊥ ⊥⊥⊥⊥	20
6	550–574	⊥⊥⊥⊥			7
7	575–599	⊥⊥⊥⊥ ⊥⊥⊥⊥ ⊥⊥⊥⊥	15
8	600–624	⊥⊥⊥⊥ ⊥⊥⊥⊥ ⊥⊥⊥⊥		16
9	625–649	⊥⊥⊥⊥ ⊥⊥⊥⊥		11
10	650–674	⊥⊥⊥⊥					9
11	675–699						4
12	700–724				2
13	725–749				2

totalled and the class frequencies written. The first three observations are 546, 689, and 490; so, as we read down the list of unclassified data, we would place a tally opposite the fifth (525–549), eleventh (675–699), and third (475–499) classes, in that order. Proceeding in like manner until we have a tally for each observation, then totalling the tally marks to obtain the class frequencies, we obtain Table 2.4.

The **cumulative frequency distribution** is a table showing the number of observations that are less than specified values. It is convenient to use the lower class limits (beginning with the second class) for the specified values. When this is done the cumulative frequencies become the frequency of the first class, sum of the frequencies of the first two classes, and so on.

The cumulative frequency distribution for the SAT-Verbal data is shown in Table 2.5.

Table 2.5

Less than	450	1
Less than	475	3
Less than	500	8
Less than	525	14
Less than	550	34
Less than	575	41
Less than	600	56
Less than	625	72
Less than	650	83
Less than	675	92
Less than	700	96
Less than	725	98
Less than	750	100

GRAPHICAL REPRESENTATION OF DATA

Now that the data have been classified, the important characteristics of the distribution of the data are much clearer. However, a pictorial representation will make the characteristics stand out even more. (Incidentally, just by examining the tallies in Table 2.4 we have a sort of graphic representation.) Besides, many people prefer data to be presented in a graphic form rather than in a numerical form. We will discuss three types of graphs which are commonly used to present data such as that in the frequency distribution in Table 2.3— the histogram, the frequency polygon, and the ogive.

Histogram. The histogram is the most common type of graph for displaying classified data. A histogram is a bar graph with no space between bars. It is drawn on a pair of co-ordinate axes, with the unit of measurement for the observations being measured along the horizontal axis and the number (or proportion) of observations being measured along the vertical axis. The vertical scale or *y*-axis is usually taken to begin at zero. The horizontal scale or *x*-axis can begin at any convenient number, and one simply selects any convenient point at which to begin the classes.

The class boundaries are marked off on the horizontal axis. The difference in value between any two successive class boundaries is represented by the width of that class interval. A rectangle, the width of whose base equals the width of the class interval and whose height is determined by the number of observations, is drawn for each class. Usually, if the class intervals are equal, the distances between the boundaries are drawn as equal. Each distance marked off on the horizontal axis may now be used as the base of a rectangle. The height of each rectangle will be determined by the class frequency—the number of observations within that class.

The histogram for the SAT-Verbal data is shown in Fig. 2.1.

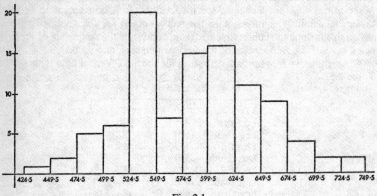

Fig. 2.1

Note that the point 424·5 is located at an arbitrary location on the horizontal axis.

An important property of a histogram is that the area of each rectangle is proportional to the number of observations in the corresponding class. In the infrequent situation where unequal class intervals are used, one needs to be careful that the areas of the rectangles are proportional to the number of observations in the classes. For example, suppose that for one reason or another we decide to combine the last three classes into a single class. The class interval of the class thus formed would be 75 units, and the class frequency would be 8. Since the class interval is three times that of the other classes, if we made the height of the rectangle 8 units the area of the last rectangle would be three times as much as it should be. It should be clear that if the class interval of a particular class is three times that of the other classes, then the rectangle corresponding to it should be only one-third as tall as the rectangle corresponding to any other class which contains the same number of observations. Thus the rectangle corresponding to the wide class we are discussing should be $\frac{8}{3}$ units high. The incorrect and correct histograms are shown in Figs. 2.2 and 2.3 respectively.

As we have already seen, the division of the range of the observations into classes is entirely arbitrary, although some hints concerning good practice can be given. The classes must be described in such a way that there is no ambiguity about the class in which each observation should be placed. The number of classes can be as small as five or six for sample sizes of twenty-five or so. Eight or ten classes can conveniently be used for samples of about fifty observations. For larger samples from ten to twenty classes will, in general, be most satisfactory. The number of classes and the class interval should be chosen so that the range of the observations is covered efficiently,

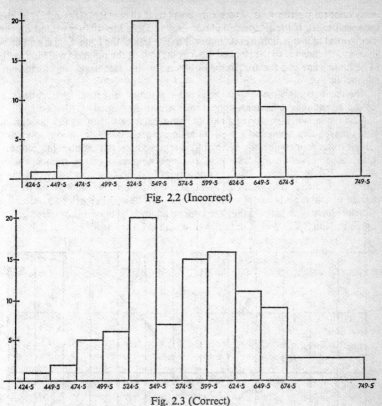

Fig. 2.2 (Incorrect)

Fig. 2.3 (Correct)

with not too much unnecessary room at the ends, and with about the same amount of 'spare room' below the smallest observation and above the largest one.

Frequency Polygon. The frequency polygon is formed by placing a dot at the mid-point, i.e. the class mark, of the top of each rectangle of the histogram

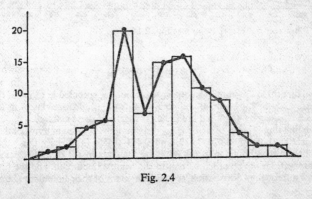

Fig. 2.4

and connecting the dots. Dots can be placed over each class mark at the proper height, if the histogram is not drawn. Dots are usually placed on the horizontal axis one-half a class interval to the left of the lowest class and one-half a class interval to the right of the highest so that the polygon will be closed. The histogram and the frequency polygon for the data under discussion are shown in Fig. 2.4.

Ogive. The graph of the cumulative frequency distribution is called an ogive. Many ogives have an appearance which is distinctly S-shaped.

The ogive is drawn on a pair of perpendicular axes, just as the histogram and the frequency polygon are, with the horizontal axis representing the values of the observations and the vertical axis representing the number (or proportion) of observations. Dots are placed opposite each of the numbers 450, 475, . . . 750, at whatever height is appropriate, to indicate how many observations are less than that value. For instance, the dot opposite 450 would be at a height of 1 unit and the dot opposite 600 would be at a height of 56. After all the dots have been located they are connected and the ogive is completed. The ogive for the SAT–Verbal data given in Table 2.5 is shown in Fig. 2.5.

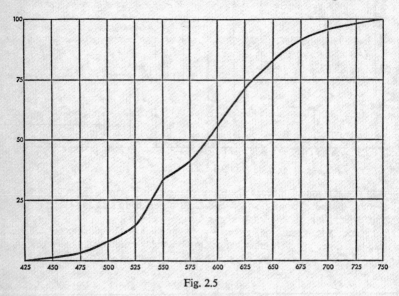

Fig. 2.5

EXERCISES

1. The first three classes of a frequency distribution, specified in terms of the class limits, are 42·5–44·4, 44·5–46·4, and 46·5–48·4. (*a*) Specify these classes in terms of their class boundaries. (*b*) What are the class marks of these classes?

2. The first three classes of a frequency distribution, specified in terms of their class limits, are 0·471–0·475, 0·476–0·480, and 0·481–0·485. (*a*) Specify these classes in terms of their class boundaries. (*b*) What are the class marks of these classes?

3. Observations are recorded correct to the nearest hundredth, and the first three classes of a frequency distribution, specified in terms of their boundaries, are 1·385–

1·425, 1·425–1·465, and 1·465–1·505. (*a*) Specify these classes in terms of their class limits. (*b*) What are the class marks of these classes?

4. (*a*) If the class marks of the first three classes of a frequency distribution are 115, 124, and 133, what are the class boundaries of these classes? (*b*) If the observations are recorded as correct to the nearest integer, what are the class limits?

5. The class marks of the first three classes of a frequency distribution are 2·475, 2·515, and 2·555. (*a*) What are the class boundaries of these classes? (*b*) What are the class limits, if the observations are recorded as correct to the nearest hundredth of a unit?

6. The class boundaries of the first three classes of a frequency distribution are given as 37·6 but not 42·1, 42·1 but not 46·6, and 46·6 but not 51·1. (*a*) What are the class marks? (*b*) What are the class limits, if the observations are given as correct to the nearest tenth of a unit? (*c*) What are the class limits, if the observations are correct to the nearest hundredth?

7. A sample contains 200 observations, recorded to the nearest tenth of a unit, ranging in value from 15·4 to 32·1. Decide how many classes you would use to classify these data. What are the boundaries of the first three classes? What are the limits of the first three classes? What are the class marks of the first three classes? [There are no unique answers.]

8. A sample consists of 34 observations, recorded as correct to the nearest integer, ranging in value from 201 to 337. If it is decided to use seven classes of width 20 units and to begin the first class at 199·5, find the class boundaries, limits, and marks of the seven classes.

9. A sample consists of 43 observations, each recorded to the nearest tenth of a unit, ranging in value from 5·1 to 13·4. If it is decided to use eight classes of width 1·1 units and to begin the first one at 4·85, find the class boundaries, limits, and marks of the eight classes.

10. The following data are the weights, correct to the nearest thousandth of a kilogram, of 27 '1 kg' packages of grapes. Classify the data. Construct the frequency distribution and the cumulative frequency distribution. Sketch the histogram, the frequency polygon, and the ogive.

1·009	1·013	0·996
1·017	0·988	1·007
0·985	0·973	1·043
1·024	1·018	1·028
1·010	0·997	1·002
0·981	1·002	1·013
1·031	0·990	0·994
1·025	1·000	1·012
1·003	1·009	1·020

CHAPTER III

MEASURES OF LOCATION

INTRODUCTION

In the previous chapter we saw that it is very difficult to learn anything by examining unordered and unclassified data. We also saw that condensing the observations into a frequency distribution aids in grasping the information that they contain. The major features of the sample data are apparent at a glance when they are classified and when the resulting frequency distribution is displayed as a histogram.

We can condense the information given in a frequency distribution still further and summarize the important information by means of just two numbers. In our study of statistics there are only two main aspects of the sample which are of interest to us. The first is the location of the data, and the various numbers that give us information about this are known as **measures of location**, or measures of **central tendency**. 'Location of the data' refers to a value which is typical of all the sample observations. Frequently a measure of location can be thought of as a measure which gives the location of the 'centre' of the data. The present chapter is concerned with the definition, illustration, and explanation of several measures of location.

The second important aspect of the data is the dispersion of the observations. By this we mean how the data are scattered (dispersed). The next chapter deals with **measures of dispersion**, also called **measures of variation**.

We will discuss four measures of location in the order of their increasing importance. The **mid-range** is seldom used, and is included primarily for the sake of completeness. The **mode**, also, is little used. The **median** and the **mean** are both used extensively. It is the author's experience that the mean is used considerably more than the median, especially in the area of testing statistical hypotheses.

THE MID-RANGE

The mid-range is the number halfway between the smallest and largest observations. By definition

$$\text{Mid-range} = \frac{\text{Smallest observation} + \text{Largest observation}}{2} \quad (3.1)$$

EXAMPLE: A sample consists of the observations 51, 47, 62, 54, 58, 65, 48, 41. The smallest observation is 41 and the largest is 65; thus we have

$$\text{Mid-range} = \frac{41 + 65}{2} = \frac{106}{2} = 53$$

EXAMPLE: For the SAT–Verbal data, we see, from inspection of the array in Table 2.2, that the smallest observation is 426 and the largest is 740. For this sample

$$\text{Mid-range} = \frac{426 + 740}{2} = \frac{1166}{2} = 583$$

THE MODE

The mode is defined as the observation in the sample which occurs most frequently, if there is such an observation. If each observation occurs the same number of times, then there is no mode. If two or more observations occur the same number of times (and more frequently than any of the other observations), then there is more than one mode, and the sample is said to be **multimodal**. If there is only one mode the sample is said to be **unimodal**.

EXAMPLE: If the sample is 14, 19, 16, 21, 18, 19, 24, 15, 19, then the mode is 19.
EXAMPLE: If the sample is 6, 7, 7, 3, 8, 5, 3, 9, then there are two modes, 3 and 7 [bimodal].
EXAMPLE: If the sample is 14, 16, 21, 19, 18, 24, 17, then there is no mode.
EXAMPLE: If the sample is the SAT–Verbal data (see Table 2.2), then the mode is 599.

THE MEDIAN

If the sample observations are arranged in order from smallest to largest, the median is defined as the middle observation if the number of observations is odd, and as the number half-way between the two middle observations if the number of observations is even.

EXAMPLE: Given the sample 34, 29, 26, 37, 31. Arranged in order we have 26, 29, 31, 34, and 37. The number of observations is odd; the median is 31.
EXAMPLE: Given the sample 34, 29, 26, 37, 31, 34. Arranged in order we have 26, 29, 31, 34, 34, 37. The number of observations is even. The median is half-way between the third and fourth (the two middle) observations. Thus the median is 32·5.
EXAMPLE: If the sample is composed of the SAT–Verbal data (see Table 2.2), then the median is h .lf-way between the fiftieth and fifty-first observations in the array; these two observations are 597 and 599. The median is 598.

THE ARITHMETIC MEAN

The most commonly used measure of location is the arithmetic mean, called simply the mean. The definition is simple:

$$\text{Sample mean} = \frac{\text{Sum of the observations}}{\text{Number of observations}} \qquad (3.2)$$

The number of observations is usually denoted by n. Also, the first (not in order of size, but simply in the order examined or written) observation is denoted x_1 (read 'x sub one' or merely 'x one'), the second observation is denoted x_2, the third is denoted x_3, and so on until the last observation, denoted x_n. The mean of the sample is denoted by the symbol $\bar{x}$ (read 'x bar'). Thus the definition above can be written

$$\bar{x} = \frac{x_1 + x_2 + x_3 + \ldots + x_n}{n} \qquad (3.3)$$

where the symbolism $\ldots + x_n$ means that we are to continue adding the observations until we reach the last one which is x_n.

EXAMPLE: If our sample consists of the data 8, 7, 11, 8, 12, 14, then the mean is

$$\bar{x} = \frac{8 + 7 + 11 + 8 + 12 + 14}{6} = \frac{60}{6} = 10$$

Note that, as the sample observations are written above, $x_1 = 8$ $x_2 = 7$. $x_3 = 11$, $x_4 = 8$, and so on.

EXAMPLE: Consider the SAT–Verbal data. From Table 2.1 we have

$$\bar{x} = \frac{x_1 + x_2 + \ldots + x_n}{n}$$

$$= \frac{546 + 689 + \ldots + 528}{100}$$

$$= \frac{58\,952}{100}$$

$$= 589 \cdot 52$$

In summary, we have found four different values for the 'centre' of the SAT–Verbal data by using four different measures of location. The mid-range equals 583, the mode equals 599, the median equals 598, and the mean equals 589·52.

To briefly compare the four measures of location discussed we make the following observations. The mid-range is easy to find, but because only two observations are involved in the definition it neglects most of the information which is present in the entire sample. The mode is a satisfactory measure of location if the frequency distribution of the sample is rather symmetrical. But if the frequency distribution is not symmetrical the most frequent observation might be far removed from the 'centre' of the sample, and the mode would not be a very good measure of location. The median has much to commend it. Its definition makes use of all observations. Extreme observations do not cause the median to fluctuate much. For instance, the median of 13, 14, 16, 18, 21 is 16 and the median of 13, 14, 16, 18, 21, 50 is 17. The 50, which is much larger than any of the other observations, causes very little change in the median. This is not true for the mean, because the mean of 13, 14, 16, 18, 21 is 16·4 and the mean of 13, 14, 16, 18, 21, 50 is 22. We have a change of about 34 per cent in the mean, which is about six times the change in the median.

The median is very easy to find when the data have been arranged in order. Another advantage of the median occurs in situations when data are classified and there is an open class. For instance, if the class intervals of a frequency distribution are 100 but not 200, 200 but not 300, 300 but not 400, 400 or more, then it is impossible to calculate the mean—because the open class, '400 or more', has no upper boundary.

One of the primary advantages of the mean as a measure of location is that if we have a mean for each of several samples and want to find the mean of the sample which results when the several samples are combined this can be easily done. If the medians of several samples are known and the median of the combined samples is desired, it cannot be found as quickly as the mean. When tests of hypotheses are made about the 'location' of the population, tests about the mean are more powerful than tests about the median, although somewhat more restrictive assumptions need to be made. Another advantage of using the mean is that the data do not have to be arranged in order, as they do when the median is used. Try Questions 1 to 6 in the Exercises, page 26.

THE MEDIAN OF CLASSIFIED DATA

Recall that the definition of the median is different for odd and for even numbers of observations when the data are not classified. However, if the n

data are classified, then the median is simply defined as the $\frac{n}{2}$th 'observation'.

Thus, if we have the frequency distribution of 100 observations, then the 50th observation in order of size would be the median; if we have 101 observations then the '50·5th' observation would be the median. If the reader is puzzled by what the phrase the ' "50·5th" observation' means, we will be in a better position to explain this phrase after the following example.

EXAMPLE: Consider the frequency distribution shown in Table 3.1:

Table 3.1

Class	Class boundaries	Frequencies
1	49·5– 99·5	17
2	99·5–149·5	38
3	149·5–199·5	61
4	199·5–249·5	73
5	249·5–299·5	56
6	299·5–349·5	29
7	349·5–399·5	16
8	399·5–449·5	10
		300

The median of this frequency distribution is the 150th observation. There are 116 observations in the first three classes, and 189 in the first four. The 150th observation is in the fourth class, which, since it contains the median, is called the median class. We know that the median lies in the interval from 199·5 to 249·5, but we don't know exactly where. We can make an estimate (an educated guess) of the median by linear interpolation if we assume that the observations are *distributed uniformly* throughout the interval from 199·5 to 249·5. The term 'distributed uniformly' means the following: if one thinks of the interval from 199·5 to 249·5 as being marked off on a scale it would be 50 units long. If the 73 observations which the class contains are distributed uniformly throughout the interval there would be an observation every $\frac{50}{73}$ of a unit. In other words, if the 50 units comprising the interval were divided into 73 equal intervals, then each interval would contain one observation.

To find the median, we must count 34 observations into the fourth class because the median is the 150th observation and the first three classes contain only 116 observations. Hence, the median is located at the point which is $\frac{34}{73}$ of the distance along the fourth class interval; this point is $\frac{34}{73}$. (50) units to the right of 199·5, the lower boundary of the fourth class. Thus, the median equals

$$199·5 + \tfrac{34}{73} . (50) = 199·5 + 23·3 = 222·8$$

We can reason in exactly the same manner to obtain a general formula for finding the median value for classified data. First, we need to find the class which contains the middle observation. Let M denote the number of this class, where M is some integer from 1 to k. If the median occurs in the fifth class, then $M = 5$; if it occurs in the seventh class, then $M = 7$; and so on. Let the frequency of the Mth class be denoted by f_M. Next, note how many observations are in the $M - 1$ classes preceding the median class; denote this cumulative frequency by F_{M-1}.

The number of observations which we must count into the median class in order to find the median equals the difference between $\frac{n}{2}$, the number of the middle observation, and the number of observations in the classes below the median class. Using the symbolism introduced in the preceding paragraph, this difference is $\frac{n}{2} - F_{M-1}$. There are f_M observations in the median class and, assuming that the observations are distributed evenly throughout the median class, the value of the median is $\dfrac{\frac{n}{2} - F_{M-1}}{f_M}$ of the distance along the class interval. Thus, for a class interval of width c units, the median is

$$\frac{\frac{n}{2} - F_{M-1}}{f_M} \cdot c$$

units to the right of the lower boundary of the median class, which we will denote by b_L. Hence, we see that the general formula for the median of classified data is

$$\text{Median} = b_L + \frac{\frac{n}{2} - F_{M-1}}{f_M} \cdot c \qquad (3.4)$$

where

b_L = lower boundary of the median class;
n = number of observations;
f_M = the number of observations in the median class; and
F_{M-1} = the number of observations in the $M - 1$ classes preceding the median class.

In the above example on Table 3.1 we had

$$b_L = 199 \cdot 5; \; n = 300; \; F_{M-1} = 116; \; f_M = 73; \; c = 50$$

So that the median $= 199 \cdot 5 + \dfrac{(150 - 116)}{73} \cdot 50$ as already indicated.

Now it should be clear why the median is defined as the $\frac{n}{2}$th observation, regardless of whether n is odd or even. Recall that in a histogram the areas of the rectangles are proportional to the numbers of observations in each of the respective classes. Thus, a vertical line through the median should divide the histogram into two parts of equal areas. This is consistent with the previous definition—that the number of observations on one side of the median equals the number of observations on the other—because the areas here are analogous to the observations in the previous definition of the median.

If we had a sample of 101 classified observations, and if we said that the median is the 51st observation we would need to find the point where the 51st observation is estimated to be. At this point an area corresponding to 51 observations is to the left, and to the right is an area corresponding to only 50 observations. Since the areas are unequal, this point is not the median. If, on the other hand, we take the 50·5th observation to be the median, then the point where this observation is estimated to be will exactly divide the histogram into two parts of equal areas—an area corresponding to 50·5 observations lies to its left, and an area corresponding to 50·5 observations lies to its right.

Relation between Measures of Location and the Types of Frequency Curves. In practice, the frequency curves have the following shapes:

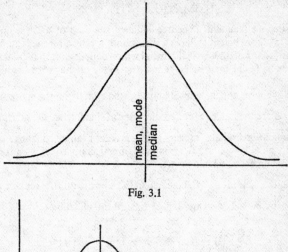

Fig. 3.1

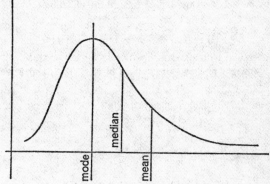

Fig. 3.2

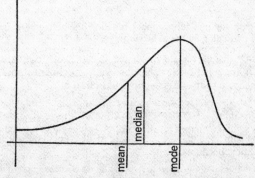

Fig. 3.3

Fig. 3.1 shows a symmetrical curve in which the mode, median, and mean coincide. Fig. 3.2 and Fig. 3.3 show two *skew* frequency curves and the relative positions of the measures of location. Notice that each one of the above curves is unimodal.

SUMMATION NOTATION

In statistics we frequently need to indicate the sum of a large number of terms, and it is convenient to have a shorthand notation for such an indicated sum. The standard mathematical symbol which denotes a sum is $\sum$, the capital Greek letter 's' (used because 's' is the first letter of the word 'sum'), called sigma.

The most general appearance of an indicated sum using the sigma-notation is $\sum_{i=m}^{n} f(i)$, where the letter i is called the **index of summation**, m and n are called **limits of summation** and are integers with n greater than or equal to m (written symbolically as $n \geq m$), and $f(i)$ is some expression involving i—a **function of** i, to use mathematical terminology.

In words, the symbol $\sum_{i=m}^{n} f(i)$ means 'in the expression $f(i)$, replace i successively by m, by $m + 1$ ..., and by n, and add the resulting terms.' Symbolically,

$$\sum_{i=m}^{n} f(i) = f(m) + f(m + 1) + \ldots + f(n) \qquad (3.5)$$

EXAMPLE: $\sum_{i=4}^{8} i$ means replace i successively by 4, 5, 6, 7, and 8 and add the resulting terms. We have

$$\sum_{i=4}^{8} i = 4 + 5 + 6 + 7 + 8 = 30$$

Similarly,

$$\sum_{i=1}^{3} 3^i = 3^1 + 3^2 + 3^3 = 39$$

and

$$\sum_{i=0}^{2} (i + 1)^i = (0 + 1)^0 + (1 + 1)^1 + (2 + 1)^2$$
$$= \quad 1 \quad + \quad 2 \quad + \quad 9$$
$$= 12$$

NOTE: Recall that any non-zero number raised to the zero power is defined to be equal to 1. In particular, in the above example, $1^0 = 1$.

Frequently in statistics we are interested in sums of observations (or of quantities which are functions of the observations). For instance, we have previously defined the sample mean (Formula (3.3)) as

$$\bar{x} = \frac{x_1 + x_2 + \ldots + x_n}{n}$$

We can write the sum $x_1 + x_2 + \ldots + x_n$ in an abbreviated fashion by using summation notation because

$$x_1 + x_2 + \ldots + x_n = \sum_{i=1}^{n} x_i$$

So, using summation notation, the definition of $\bar{x}$ becomes

$$\bar{x} = \frac{\sum\limits_{i=1}^{n} x_i}{n} \tag{3.6}$$

Some general properties of the summations of various quantities are derived below. We will refer to these properties occasionally throughout the book, so you should become familiar with them.

If the expression $f(i)$ does not contain an i it is called a **constant function**, or more simply, a **constant**. So $f(i) = 3$, $f(i) = 15a$, and $f(i) = c$ are examples of constant functions. By our rule, the symbol $\sum\limits_{i=1}^{n} c$ means that in the expression $f(i) = c$ we should replace i by 1, i by 2, and so on, until finally we replace i by n; then we are to add the resulting terms. When we replace i by 1 in the expression $f(i) = c$ we get $f(1) = c$; when we replace i by 2 we get $f(2) = c$; and so on. So we have

$$\sum_{i=1}^{n} c = \overbrace{c + c + \ldots + c}^{n \text{ terms}} = nc \tag{3.7}$$

Stated verbally, when we sum a constant from $i = 1$ to $i = n$ we get n times that constant. More generally,

$$\sum_{i=m}^{n} c = (n - m + 1)c \tag{3.8}$$

EXAMPLES:
$$\sum_{i=0}^{5} 4 = 4 + 4 + 4 + 4 + 4 + 4 = 24$$

$$\sum_{i=3}^{5} 6 = 6 + 6 + 6 = 18$$

$$\sum_{i=1}^{100} c = 100c$$

Next we will show that $\sum\limits_{i=1}^{n} ax_i = a \sum\limits_{i=1}^{n} x_i$. By definition we have

$$\sum_{i=1}^{n} ax_i = ax_1 + ax_2 + \ldots + ax_n,$$

and by elementary algebra we have

$$ax_1 + \ldots + ax_n = a(x_1 + x_2 + \ldots + x_n)$$

But, by definition, $\quad x_1 + x_2 + \ldots + x_n = \sum\limits_{i=1}^{n} x_i$

Therefore we have shown that

$$\sum_{i=1}^{n} ax_i = a \sum_{i=1}^{n} x_i$$

We frequently need to work with the summation of expressions containing two or more terms. For instance, consider the expression $\sum\limits_{i=1}^{n} (x_i + y_i)$. By definition, we have

$$\sum_{i=1}^{n} (x_i + y_i) = (x_1 + y_1) + (x_2 + y_2) + \ldots + (x_n + y_n) \quad (3.9)$$

Regrouping the quantities on the right-hand side, Equation (3.9) becomes

$$\sum_{i=1}^{n} (x_i + y_i) = (x_1 + \ldots + x_n) + (y_1 + \ldots + y_n) \quad (3.10)$$

Applying the definition to each of the two expressions on the right-hand side of the previous equation, we have

$$\sum_{i=1}^{n} (x_i + y_i) = \sum_{i=1}^{n} x_i + \sum_{i=1}^{n} y_i \quad (3.11)$$

This result generalizes, of course, to sums such as

$$\sum_{i=1}^{n} (x_i + y_i + z_i) \quad \text{or} \quad \sum_{i=1}^{n} (x_i + y_i + z_i + w_i)$$

A final expression which is rather common (we will work with this one frequently in the mean of classified data which will be discussed a little later in this chapter, and in the variance of classified data in the next chapter) is $\sum\limits_{i=1}^{n} (ax_i + b)$. By definition

$$\sum_{i=1}^{n} (ax_i + b) = (ax_1 + b) + \ldots + (ax_n + b) \quad (3.12)$$

which, after regrouping the right-hand side, becomes

$$\sum_{i=1}^{n} (ax_i + b) = (ax_1 + ax_2 + \ldots + ax_n) + (b + \ldots + b) \quad (3.13)$$

which yields

$$\sum_{i=1}^{n} (ax_i + b) = a(x_1 + \ldots + x_n) + nb \quad (3.14)$$

and finally

$$\sum_{i=1}^{n} (ax_i + b) = a \sum_{i=1}^{n} x_i + nb \quad (3.15)$$

THE MEAN OF CLASSIFIED DATA

We have previously discussed finding the mean of a sample. If there are very many observations, then the addition involved in finding $\sum\limits_{i=1}^{n} x_i$ becomes tedious, and it is advantageous to classify the data before finding the mean.

When data are classified the individual observations lose their identity. Either the experimenter is no longer interested in the values of the original observations, no longer has a record of them, or has never seen them because someone supplied the data after they had been classified.

In order to calculate the mean (and also the variance, treated in the next chapter) of the data, each observation in a class is assumed to have a value equal to that of the mid-point of the class which it occupies. If there are f_1 observations in the first class the total value of the observations in the first class is the product $x_1' . f_1$ where x_1' denotes the class mark. The total value of the observations in the second class would be the product $x_2' . f_2$, and so on.

If the data are classified into k classes the sum of the observations would be

$$x_1'f_1 + x_2'f_2 + \ldots + x_k'f_k$$

which can be written using summation notation as $\sum_{i=1}^{k} x_i'f_i$. The total number of observations would be the sum of the class frequencies and is denoted by n, as before; symbolically, $f_1 + f_2 + \ldots + f_k = n$. Recalling that the sample mean has been defined as the sum of the observations divided by the number of observations, we have

$$\bar{x} = \frac{\sum_{i=1}^{k} x_i'f_i}{n} \qquad (3.16)$$

as our definition of the mean of classified data,

where
$x_i' = $ class mark of ith class;
$f_i = $ number of observations in ith class;
$k = $ number of classes; and
$n = $ total number of observations.

EXAMPLE: Find the mean of the data given in the frequency distribution in Table 3.2, which follows:

Table 3.2

Class	Class mark	Frequency	
	x_i'	f_i	$x_i'f_i$
1	46	4	184
2	51	1	51
3	56	2	112
4	61	2	122
5	66	2	132
6	71	9	639
7	76	5	380
8	81	10	810
9	86	4	344
10	91	8	728
11	96	3	288
		50	$3,790 = \sum_{i=1}^{11} x_i'f_i$

From the definition, $\bar{x} = \dfrac{\sum_{i=1}^{11} x_i'f_i}{n} = \dfrac{3790}{50} = 75 \cdot 8$.

If the computation of a sample mean is being done with paper and pencil (that is, a desk calculator is not being used), then the arrangement shown in Table 3.2 is perhaps the most efficient and convenient—a column for the class marks, a column for the frequencies, and one for the products of class marks by frequencies. (Although columns for class boundaries and tallies are desirable when the data are being classified, they are no longer needed when data which have already been classified are being presented.)

For samples about the size of the one here (fifty observations) the computation would not be excessively troublesome if the data were added without classifying. In fact, the mean can probably be found faster when the data are not classified than when they are—it is the classification which takes time, not the computation—if the data are no more numerous than in the present example. But suppose that one has a sample of 200, or 500, or 1,000 observations? Then the classification method is without rival. Also classifying the data is advantageous, even for samples as small as forty or fifty observations, if any quantity other than the sample mean must be computed (for instance, the sample variance, which is treated in the next chapter). Generally speaking, if a desk calculator is used in most cases it is not worthwhile to classify the data, because the necessary computations can be performed so quickly, even for samples of size 100 or so.

EXERCISES

1. Find the mid-range of each of the following samples:

(a) 5, 10, 11, 6, 13, 10, 8   (c) 1·672, 1·541, 1·603, 1·659, 1·499, 1·591, 1·630
(b) 31, 42, 37, 55, 70, 52   (d) 101·4, 150·1, 134·8, 139·2, 124·3

2. Find the mode of each of the following samples:

(a) 5, 8, 11, 9, 8, 6, 8   (c) 42·57, 51·83, 47·34, 49·05
(b) 7, 9, 10, 9, 12, 11, 10, 13   (d) 7, 12, 8, 7, 10, 11, 8, 6, 10, 13, 7, 8

3. Find the median of each of the samples in Exercise 1.

4. The nine male students in a certain class received the following scores on a short quiz (a score of 50 was perfect). 48, 36, 33, 39, 30, 47, 35, 41, 38. Find the median.

5. Find the median of the sample 37, 35, 40, 35, 33, 36, 35.

6. Find the mean for each of the samples in Exercise 1.

7. Find the mean of the following data.

395	369	374	348	373
376	348	360	386	377
372	337	378	359	351
367	376	380	368	382

8. Classify the following data and find the median and the mean of the resulting classified data:

43·0	37·0	43·8	48·7	48·5
41·0	58·4	51·4	42·6	40·9
40·3	38·0	44·2	58·4	46·6
43·6	36·7	53·7	52·7	53·0
38·4	48·6	44·4	39·6	53·4
45·9	44·1	46·5	38·0	50·3
37·0	61·8	33·8	56·6	57·7
42·3	46·0	46·8	38·4	39·3
35·9	47·5	58·1	37·9	47·0
52·4	49·1	50·2	50·2	43·4
62·0	54·0	54·7	60·2	52·5
37·3	46·8	49·5	40·0	

9. Find the median and the mean of the data that were classified in Exercise 10 of Chapter II.

CHAPTER IV

MEASURES OF VARIATION

INTRODUCTION

We noted in the introduction to Chapter III that it is difficult to grasp the significant aspects of a mass of unordered sample data, and that ordering the data, classifying them to form a frequency distribution, and displaying the resulting frequency distribution graphically as a histogram are successive improvements in describing the data. However, mere pictorial descriptions are seldom adequate in making comparisons or decisions in a scientific manner. In order to compare samples in a quantitative way, the data should be described by means of numbers. Chapter III was concerned with certain quantities (mean, median, mode, and mid-range) which give numbers that are used as measures of location, supplying condensed information about where the centre of the sample is.

Consider the following three samples and their means:

Sample 1:   66, 66, 66, 67, 67, 67, 68, 69. $\bar{x} = 67$
Sample 2:   52, 53, 61, 67, 71, 72, 78, 82. $\bar{x} = 67$
Sample 3:   43, 44, 50, 54, 67, 90, 91, 97. $\bar{x} = 67$

Each of these samples has a mean equal to 67. However, the dispersion of the observations in the three samples differ greatly. (By 'dispersion of the observations in a sample' we mean the way in which the observations are spread out.) In the first sample all observations are grouped within 2 units of the mean. In the second only one observation (67) is closer to the mean than 4 units and some are as many as 15 units away. Only one observation (67) is closer than 13 units to the mean of the third sample, and some are as far away as 30 units.

Yet, when we describe each of these samples by means of a single number— a number that gives us information about where the 'centre' of the sample is— the samples are each described by the same value, namely 67. Clearly, if samples as different as the three above have the same mean, then we need to describe a sample in other ways as well. In particular, it would be desirable if we had some single numerical measure that would indicate how dispersed the data are. If we have information about where the data are located, and about the manner in which they are spread out, then these two characteristics together tell us a great deal more about the sample than does a single measure giving information only about the location of the data. Several quantities that are used as **measures of dispersion** are the **range**, the **mean absolute deviation**, the **variance**, and the **standard deviation**. These quantities are also called **measures of variation**, and the two terms 'measure of dispersion' and 'measure of variation' will be used interchangeably in this book.

Samples composed of classified data might also have the same mean, even though their frequency distributions are vastly different. For example, the samples whose histograms are shown in Figs. 4.1, 4.2, and 4.3 have the same mean.

27

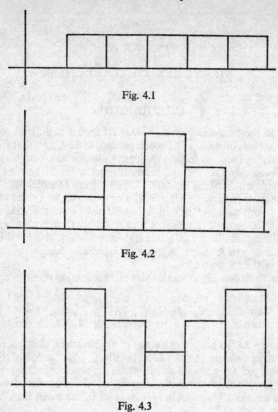

Fig. 4.1

Fig. 4.2

Fig. 4.3

THE RANGE

The range is the simplest measure of dispersion. By definition, the range equals the largest sample observation minus the smallest sample observation. The range is especially easy to find after the data have been arranged in order —one merely notes the largest and smallest observations, and finds the difference between them. If the data have not been numerically ordered they must be scanned for the largest and smallest observations. One major objection to the range is that it does not make use of all of the observations in the sample (thereby disregarding much available information), but uses only two of the observations—the largest and the smallest.

EXAMPLE: For the sample observations 13, 23, 11, 17, 25, 18, 14, 24 we see that the largest observation is 25 and the smallest observation is 11. The range is $25 - 11 = 14$.

EXAMPLE: A sample is composed of the observations 67, 79, 87, 93, 97, 57, 44, 80, 47, 78, 81, 90, 65, 88, 91. The largest observation is 97; the smallest observation is 44. The range is $97 - 44 = 53$.

EXAMPLE: For the sample data in Table 2.2 we see that the range is $740 - 426 = 314$.

THE MEAN ABSOLUTE DEVIATION

The mean absolute deviation is defined exactly as the words indicate. The word 'deviation' refers to the deviation of each observation from the mean of the sample. The term 'absolute deviation' means the numerical (i.e. positive) value of the deviation, and the 'mean absolute deviation' is simply the arithmetic mean of the absolute deviations. As our definition for the mean absolute deviation, denoted by M.A.D., we have

$$\text{M.A.D.} = \frac{\sum_{i=1}^{n} |x_i - \bar{x}|}{n} \tag{4.1}$$

Some texts refer to this as the 'mean deviation' or 'average deviation'.

Before giving examples of the calculation of the M.A.D. we must digress briefly to discuss absolute values, which might be somewhat unfamiliar. The symbolism $|x_i - \bar{x}|$ denotes the absolute value of the quantity $(x_i - \bar{x})$. The absolute value of a number is simply the value of that number without regard to its sign (that is, disregarding a negative sign).

More mathematically, we define the absolute value of any number, denoted y, as follows:

$$|y| = y \text{ if } y \text{ is positive or zero}$$
$$|y| = -y \text{ if } y \text{ is negative}$$

This formulates precisely the idea expressed loosely in the previous paragraph. Thus, the absolute value of any number is the number itself if that number is not negative. If the number is negative, then the absolute value of that number is the negative of it (which is positive). Consider the following illustrations of the application of the definition of absolute values.

EXAMPLE: 1: Find $|4|$. $|4| = 4$ (4 is not negative)
EXAMPLE: 2: Find $|0|$. $|0| = 0$ (0 is not negative)
EXAMPLE: 3: Find $|-4|$. $|-4| = -(-4) = 4$ (−4 is negative)

Now that we know how to find absolute values we can return to the discussion of mean absolute deviations.

EXAMPLE: Suppose that our sample consists of the observations 21, 17, 13, 25, 9, 19, 6, and 10. The sample mean is

$$\bar{x} = \frac{\sum_{i=1}^{n} x_i}{n} = \frac{120}{8} = 15$$

After $\bar{x}$ has been found, the sequence of computations is to:

(1) Find and record the signed differences.
(2) Find and record the absolute differences.
(3) Find $\sum_{i=1}^{n} |x_i - \bar{x}|$.
(4) Find the mean absolute deviation.

[Step (1) can be omitted, of course, and the absolute differences recorded without any intermediate steps.]

Perhaps the best manner to display the computations in steps (1), (2), and (3) is to make use of a table composed of three columns. The sample observations are written in the first column, the signed differences $(x_i - \bar{x})$ are written in the second, and the absolute differences $|x_i - \bar{x}|$ are written in the third. Then $\sum_{i=1}^{n} |x_i - \bar{x}|$ can be found simply by adding the entries in the last column. The computations indicated in steps (1), (2), and (3) are shown in Table 4.1, and the computation of the M.A.D. is shown immediately following the table.

Table 4.1

x_i	$x_i - \bar{x}$	$\|x_i - \bar{x}\|$
21	6	6
17	2	2
13	−2	2
25	10	10
9	−6	6
19	4	4
6	−9	9
10	−5	5
120		$44 = \sum_{i=1}^{8} \|x_i - \bar{x}\|$

$$\text{M.A.D.} = \frac{\sum_{i=1}^{n} |x_i - \bar{x}|}{n} = \frac{44}{8} = 5 \cdot 5$$

Then, on the average, each observation is 5·5 units from the sample mean.

The mean absolute deviation is an easy measure of dispersion to find, is simple to understand and interpret, and uses all of the observations. However, it does not yield any further elegant mathematical statistical results, as does the variance (the measure discussed in the next section), because the absolute values are rather unsuitable for mathematical analysis. For instance, suppose that several samples have been drawn from the same population (or from several populations which are assumed to have the same unknown variance). Then there are several different sample variances available, each of which is an estimate of the same population variance. These several estimates can be combined in a certain manner (which will be treated later) to give what is known as a pooled estimate of the population variance, which will be a more accurate estimate than any of the sample variances taken singly. However, if several mean absolute deviations are available from the same several samples there is no quick way in which they can be combined to give a pooled estimate of the M.A.D. for the combined samples. The M.A.D. for the combined samples can be found only by lengthy application of the definition of the M.A.D. after a single large sample has been formed from the several smaller ones.

THE VARIANCE AND THE STANDARD DEVIATION

From the remarks in the previous section it follows that it would be desirable to have a measure of variation that does not involve absolute values.

Instead of a mean absolute deviation, it might occur to us to define a 'mean signed deviation', such as

$$\text{'Mean signed deviation'} = \frac{\sum_{i=1}^{n} (x_i - \bar{x})}{n}$$

The only trouble with such an attempted definition is that it would not give us much information about the variation present in the data; the 'mean signed deviation' would be zero for every sample because the sum $\sum_{i=1}^{n} (x_i - \bar{x})$ equals zero for every sample.

It seems, then, that the consideration of signed deviations is not a fruitful approach for defining a measure of variation. We would like to deal with non-negative quantities, although we do not want to have to deal with absolute values. Thus, we want to eat our mathematical cake and have it, too. We can do this if, rather than taking the absolute value of each deviation, we square each one. This assures that each quantity will be non-negative, and avoids the necessity of having to work with absolute values.

In order to obtain a measure of variation that is the average of these squared deviations of the sample observations from the sample mean, we find the sum of the n-squared deviation, and then divide this sum by n. The squared deviations of the observations x_1, x_2, . . ., x_n from the sample mean are $(x_1 - \bar{x})^2$, $(x_2 - \bar{x})^2$, . . ., $(x_n - \bar{x})^2$. The sum of these squared deviations is

$$(x_1 - \bar{x})^2 + (x_2 - \bar{x})^2 + \ldots + (x_n - \bar{x})^2$$

which can be neatly written, using summation notation, as $\sum_{i=1}^{n} (x_i - \bar{x})^2$. A sum of squared deviations is usually called simply a **sum of squares**. Thus, whenever the term 'sum of squares' is used in statistics it is understood that it refers to the sum of squares of the deviations of the sample observations from their mean.

In order to obtain a measure of variation that is the average of the squared deviations of the n sample observations from the sample mean, we must divide the sum of squares by n. The resulting quantity is denoted by s^2 and is called the **sample mean squared deviation** or, more usually, the **sample variance**; symbolically,

$$s^2 = \frac{\sum_{i=1}^{n} (x_i - \bar{x})^2}{n} \tag{4.2}$$

For theoretical reasons, the sum of squares is usually divided by $n - 1$ rather than by n to give an 'average' deviation simply because the result obtained represents a better estimate of the 'standard deviation', which is defined below. For $n > 35$ there is practically no difference in the definitions. In this book we will use $n - 1$ as the divisor of the sum of squares of deviations, and our definition of the sample variance will be given by

$$s^2 = \frac{\sum_{i=1}^{n} (x_i - \bar{x})^2}{n - 1}, \tag{4.3}$$

rather than by Equation (4.2). In many texts, however, the sample variance is defined by Equation (4.2).

The original observations are measured in units; the deviations $(x_i - \bar{x})$ are also measured in units; hence, the squared deviations $(x_i - \bar{x})^2$ are given in terms of squared units. Since the definition of the variance involves the quantities $(x_i - \bar{x})^2$, the variance is given in terms of squared units also. For instance, if our sample data are the heights of randomly selected stalks of corn, measured in centimetres, the sample variance would be in terms of cm². It is frequently desirable to have a measure of dispersion whose units are the same as those of the observations. Since the variance is given in squared units, the square root of the variance would be given in units. Thus, if we take the square root of the variance, we have the measure of dispersion that is known as the **sample standard deviation** and denoted by s. By definition we have

$$s = \sqrt{\frac{\sum_{i=1}^{n} (x_i - \bar{x})^2}{n - 1}} \qquad (4.4)$$

EXAMPLE: Find the variance and the standard deviation for the sample data 21, 17, 13, 25, 9, 19, 6, and 10. (These are the same data as in the first example in the previous section.) Use the definition of s^2, of **4.3**.

When we compute s^2 by applying Formula (4.3), the computations can most conveniently be shown in a table. The table will be composed of three columns: a column for the observations x_i, a column for the deviations of the observations from the sample mean $(x_i - \bar{x})$, and a column for the squared deviations $(x_i - \bar{x})^2$. The first two columns are the same as those in Table 4.1, which contained computations for

Table 4.2

x_i	$x_i - \bar{x}$	$(x_i - \bar{x})^2$
21	6	36
17	2	4
13	−2	4
25	10	100
9	−6	36
19	4	16
6	−9	81
10	−5	25

$$302 = \sum_{i=1}^{8} (x_i - \bar{x})^2$$

$$s^2 = \frac{\sum_{i=1}^{8} (x_i - \bar{x})^2}{n - 1} = \frac{302}{7} = 43 \cdot 14$$

$$s = \sqrt{s^2} = \sqrt{43 \cdot 14} = 6 \cdot 57$$

Using **4.2** we obtain

$$s^2 = \frac{302}{8} = 37 \cdot 75$$

$$\therefore \quad s = 6 \cdot 15$$

the mean absolute deviation. In order to find $\sum_{i=1}^{n} (x_i - \bar{x})^2$ all we need do is to find the sum of the elements in the third column. Then we divide the sum $\sum_{i=1}^{n} (x_i - \bar{x})^2$ by $n - 1$ in order to find s^2. Note that the calculation of s^2 is similar to the computation of the mean absolute deviation. The calculations necessary to compute s^2 and s from their definitions are shown in Table 4.2.

When the number of observations is small it does not take long to compute s^2 from its definition, especially if all observations and the mean are integral values (or are rounded to integral values), as in the above example. If, however, the number of observations is large (assuming that the data are unclassified), the computation necessary to find s^2 from the definition is rather laborious, and a desk calculator would ordinarily be used. We derive below a computing formula for s^2 that is especially well-adapted to the use of a desk caculator.

We begin the derivation of the computing formula for s^2 with the definition of s^2,

$$s^2 = \frac{1}{n-1} \sum_{i=1}^{n} (x_i - \bar{x})^2 \qquad (4.5)$$

Expanding the quantity $(x_i - \bar{x})^2$, we get

$$s^2 = \frac{1}{n-1} \sum_{i=1}^{n} (x_i^2 - 2x_i\bar{x} + \bar{x}^2) \qquad (4.6)$$

which, since the summation of the sum of several terms is the sum of the summations of the separate terms, becomes

$$s^2 = \frac{1}{n-1} \left[\sum_{i=1}^{n} x_i^2 - \sum_{i=1}^{n} 2x_i\bar{x} + \sum_{i=1}^{n} \bar{x}^2 \right] \qquad (4.7)$$

Using the facts that $\sum_{i=1}^{n} 2x_i\bar{x} = 2\bar{x} \sum_{i=1}^{n} x_i$ and $\sum_{i=1}^{n} \bar{x}^2 = n\bar{x}^2$, both of which follow because 2 and $\bar{x}$ are constants, we obtain

$$s^2 = \frac{1}{n-1} \left[\sum_{i=1}^{n} x_i^2 - 2\bar{x} \sum_{i=1}^{n} x_i + n\bar{x}^2 \right] \qquad (4.8)$$

Replacing $\bar{x}$ by $\dfrac{\sum_{i=1}^{n} x_i}{n}$, which it equals by definition, we obtain

$$s^2 = \frac{1}{n-1} \left[\sum_{i=1}^{n} x_i^2 - 2\left(\frac{\sum_{i=1}^{n} x_i}{n}\right)\left(\sum_{i=1}^{n} x_i\right) + n\left(\frac{\sum_{i=1}^{n} x_i}{n}\right)^2 \right] \qquad (4.9)$$

which, after some algebra, becomes

$$s^2 = \frac{1}{n-1} \left[\sum_{i=1}^{n} x_i^2 - \frac{2\left(\sum_{i=1}^{n} x_i\right)^2}{n} + \frac{\left(\sum_{i=1}^{n} x_i\right)^2}{n} \right] \qquad (4.10)$$

Combining the last two terms within the brackets in Formula (4.10), we finally obtain

$$s^2 = \frac{1}{n-1}\left[\sum_{i=1}^{n} x_i^2 - \frac{\left(\sum_{i=1}^{n} x_i\right)^2}{n}\right] \qquad (4.11)$$

This formula is valuable because the two sums, $\sum_{i=1}^{n} x_i$ and $\sum_{i=1}^{n} x_i^2$, can both be found quickly and simultaneously on a desk calculator. It will nearly always shorten 'pencil and paper' computations somewhat, and frequently the computational work necessary will be considerably less.

In Equation (4.8) if we had multiplied and divided the middle term within the square brackets by n we would have had

$$s^2 = \frac{1}{n-1}\left[\sum_{i=1}^{n} x_i^2 - 2n\bar{x}\left(\frac{\sum_{i=1}^{n} x_i}{n}\right) + n\bar{x}^2\right] \qquad (4.12)$$

After replacement of $\dfrac{\sum_{i=1}^{n} x_i}{n}$ by $\bar{x}$, Equation (4.12) becomes

$$s^2 = \frac{1}{n-1}\left[\sum_{i=1}^{n} x_i^2 - 2n\bar{x}^2 + n\bar{x}^2\right] \qquad (4.13)$$

and finally $$s^2 = \frac{1}{n-1}\left[\sum_{i=1}^{n} x_i^2 - n\bar{x}^2\right] \qquad (4.14)$$

Formula (4.14) is an alternative computing formula that is occasionally used to find s^2, especially if a value for $\bar{x}$ has already been calculated.

EXAMPLE: Using the same sample as in the previous example (21, 17, 13, 25, 9, 19, 6, and 10), find s^2 by using the computing formula given in Formula (4:14).

The observations, x_i, and their squares, x_i^2 are displayed in Table 4.3 below. To find the necessary sums $\sum_{i=1}^{n} x_i$ and $\sum_{i=1}^{n} x_i^2$, simply add the entries in each column

Table 4.3

x_i	x_i^2
21	441
17	289
13	169
25	625
9	81
19	361
6	36
10	100
$\sum_{i=1}^{8} x_i = 120$	$2102 = \sum_{i=1}^{8} x_i^2$

We have $\sum_{i=1}^{n} x_i = 120$, $\sum_{i=1}^{n} x_i^2 = 2102$, and $n = 8$. Then from its definition we have

$$\bar{x} = \frac{\sum_{i=1}^{n} x_i}{n} = \frac{120}{8} = 15$$

Substituting the appropriate values for n, $\sum_{i=1}^{n} x_i^2$, and $\bar{x}$ into Formula (4.14), we have

$$s^2 = \frac{1}{n-1}\left[\sum_{i=1}^{n} x_i^2 - n\bar{x}^2\right]$$

$$= \frac{1}{8-1}[2102 - 8(15)^2]$$

$$= \frac{1}{7}[2102 - 1800]$$

$$= \frac{302}{7}$$

$$s^2 = 43 \cdot 14$$

Also, $\qquad\qquad s = \sqrt{43 \cdot 14} = 6 \cdot 57$

Note that these values of s^2 and s agree with those which we obtained in the previous example in this section. This is to be expected, because the computing formula that we used is algebraically equivalent to the definition.

THE VARIANCE AND STANDARD DEVIATION OF CLASSIFIED DATA

If the data have been classified into k classes, then by definition

$$s^2 = \frac{1}{n-1}\sum_{i=1}^{k} (x_i' - \bar{x})^2 f_i \qquad (4.15)$$

where x_i' = the midpoint (class mark) of the ith class;

f_i = the number of observations in the ith class;

n = the total number of observations: $n = \sum_{i=1}^{k} f_i$; and

$\bar{x}$ = the sample mean: $\bar{x} = \frac{1}{n}\sum_{i=1}^{k} x_i' f_i$.

The sample variance can be computed from its definition, although a computing formula for s^2 is usually employed when a desk calculator is being used. Such a computing formula will be derived after the next example.

In the example below s^2 is found by a straightforward application of the definition in Formula (4.15). The numbers have been deliberately chosen so that the computations will be easy.

EXAMPLE: A history test was taken by 51 students. The scores ranged from 50 to 95 and were classified into 8 classes of width 6 units. The resulting frequency distribution appears in Table 4.4. Find s^2 by applying the definition for s^2. Then find s.

36   *Statistics Made Simple*

Table 4.4

Class i	Class Mark x_i'	Frequency f_i	$x_i'f_i$
1	51	2	102
2	57	3	171
3	63	5	315
4	69	8	552
5	75	10	750
6	81	12	972
7	87	10	870
8	93	1	93
		51	$3825 = \sum_{i=1}^{8} x_i'f_i$

We must first find $\bar{x}$: $\quad \bar{x} = \dfrac{\sum_{i=1}^{8} x_i'f_i}{n} = \dfrac{3825}{51} = 75$

Now we need to find $\sum_{i=1}^{k} (x_i' - \bar{x})^2 f_i$. In order to find the products $(x_i' - \bar{x})^2 f_i$ we must first find the squared quantities $(x_i' - \bar{x})^2$. We need three columns to display the computation of the quantities $(x_i' - \bar{x})^2$—a column for the x_i', a column for the $(x_i' - \bar{x})$, and a column for the $(x_i' - \bar{x})^2$. We also need a column for f_i and a final column for the products $(x_i' - \bar{x})^2 f_i$. The necessary computations for finding $\sum_{i=1}^{k} (x_i' - \bar{x})^2 f_i$ are shown in Table 4.5.

Table 4.5

Class	x_i'	f_i	$x_i' - \bar{x}$	$(x_i' - \bar{x})^2$	$(x_i' - \bar{x})^2 f_i$
1	51	2	−24	576	1152
2	57	3	−18	324	972
3	63	5	−12	144	720
4	69	8	−6	36	288
5	75	10	0	0	0
6	81	12	6	36	432
7	87	10	12	144	1440
8	93	1	18	324	324

$$\sum_{i=1}^{8} f_i = 51 \qquad\qquad 5328 = \sum_{i=1}^{8} (x_i' - \bar{x})^2 f_i$$

Thus we have $\quad s^2 = \dfrac{1}{n-1} \sum_{i=1}^{8} (x_i' - \bar{x})^2 f_i = \dfrac{1}{50}(5328) = 106\cdot56$

$$s = \sqrt{106\cdot56} = 10\cdot323$$

We now proceed to derive a computing formula (and also an alternative computing formula) for the sample variance for classified data. Squaring $(x_i' - \bar{x})$ in Formula (4.15), and then multiplying by f_i, we obtain

$$s^2 = \frac{1}{n-1} \sum_{i=1}^{k} (x_i'^2 f_i - 2x_i'\bar{x}f_i + \bar{x}^2 f_i) \qquad (4.16)$$

which becomes

$$s^2 = \frac{1}{n-1}\left[\sum_{i=1}^{k} x_i'^2 f_i - 2\bar{x}\sum_{i=1}^{k} x_i' f_i + \bar{x}^2 \sum_{i=1}^{k} f_i\right] \quad (4.17)$$

Recall that $\sum_{i=1}^{k} f_i = n$ and that, by definition, $\bar{x} = \dfrac{\sum_{i=1}^{k} x_i f_i}{n}$. Upon substitution

of these values for $\sum_{i=1}^{k} f_i$ and $\bar{x}$, Equation (4.17) becomes

$$s^2 = \frac{1}{n-1}\left[\sum_{i=1}^{k} x_i'^2 f_i - 2\frac{\left(\sum_{i=1}^{k} x_i' f_i\right)^2}{n} + \frac{\left(\sum_{i=1}^{k} x_i' f_i\right)^2}{n}\right] \quad (4.18)$$

Combining the second and third terms within brackets in Equation (4.19), we obtain

$$s^2 = \frac{1}{n-1}\left[\sum_{i=1}^{k} x_i'^2 f_i - \frac{\left(\sum_{i=1}^{k} x_i' f_i\right)^2}{n}\right] \quad (4.19)$$

In the middle term within brackets in Equation (4.18) if we had replaced $\sum_{i=1}^{k} x_i' f_i$ by $n\bar{x}$ (it follows from the definition of $\bar{x}$ for classified data that these two quantities are equal), we would have had

$$s^2 = \frac{1}{n-1}\left[\sum_{i=1}^{k} x_i'^2 f_i - 2\bar{x}\cdot n\bar{x} + \bar{x}^2 \cdot n\right] \quad (4.20)$$

which, upon combining the two last terms within the brackets, yields

$$s^2 = \frac{1}{n-1}\left[\sum_{i=1}^{k} x_i'^2 f_i - n\bar{x}^2\right] \quad (4.21)$$

Note that Equation (4.19) is analogous to Equation (4.11) and that Equation (4.21) is analogous to Equation (4.14).

The following example illustrates the computation of the sample variance for classified data by means of the computing formula given in Equation (4.20). The 50 sample observations are test scores in a freshman mathematics class.

EXAMPLE: The data in Table 4.6 are the scores of 50 students on a mathematics test. Classify the data, and then find the sample variance by applying the computing formula (4.19).

In order to classify the data we must first decide how many classes would be appropriate. The range is 54 units (Range $= 98 - 44 = 54$). Then eleven classes, each of width 5 units, would be sufficient to cover the range of the sample. (Among various other possibilities, fourteen classes of width 4 units each or eight classes of width 7 units each would also be satisfactory.) The lower boundary of the first class we will choose as 43·5; then succeeding class boundaries will be 48·5, 53·5, 58·5, and so on to the upper boundary of the last class, which will be 98·5. These boundaries will give convenient integral values for the class marks and, since they have one more decimal place of accuracy than the observations themselves, we know that no observation will equal a class boundary.

Table 4.6

67	92	98	49	80
45	93	80	81	72
95	97	92	78	76
79	72	84	69	88
70	60	74	74	83
82	57	47	90	70
83	48	60	71	71
84	86	57	82	91
71	44	78	73	90
93	93	79	65	83

The computations necessary for finding the sums $\sum_{i=1}^{k} x_i' f_i$ and $\sum_{i=1}^{k} x_i'^2 f_i$ are shown in Table 4.7. When a desk calculator is being used it is unnecessary to write down the products $x_i' f_i$ and $x_i'^2 f_i$; these products would be accumulated on the calculator. To display the relevant quantities involved in computing the products $x_i' f_i$ we need columns for the x_i' and the f_i and a column for the resulting $x_i' f_i$. In order to display the relevant quantities involved in computing the $x_i'^2 f_i$ we will need a column for the $x_i'^2$, in addition to the f_i column which we already have, and a column for the resulting $x_i'^2 f_i$.

Thus, we need five columns for the quantities used in the computation of $\sum_{i=1}^{k} x_i' f_i$ and $\sum_{i=1}^{k} x_i'^2 f_i$—one for each of the quantities $x_i', f_i', x_i' f_i, x_i'^2, x_i'^2 f_i$. The necessary sums $\sum_{i=1}^{k} f_i$, $\sum_{i=1}^{k} x_i' f_i$, and $\sum_{i=1}^{k} x_i'^2 f_i$ are found by adding the entries in the appropriate columns. In addition to these five 'computation' columns, Table 4.8 contains two 'information' columns, one for the numbers of the classes and another for the class boundaries.

The remaining computations for s^2, for the data in Table 4.7, are shown following Table 4.7.

Table 4.7

Class	Class boundaries	x_i'	f_i	$x_i'^2$	$x_i' f_i$	$x_i'^2 f_i$
1	43·5–48·5	46	4	2 116	184	8 464
2	48·5–53·5	51	1	2 601	51	2 601
3	53·5–58·5	56	2	3 136	112	6 272
4	58·5–63·5	61	2	3 721	122	7 442
5	63·5–68·5	66	2	4 356	132	8 712
6	68·5–73·5	71	9	5 041	639	45 369
7	73·5–78·5	76	5	5 776	380	28 880
8	78·5–83·5	81	10	6 561	810	65 610
9	83·5–88·5	86	4	7 396	344	29 584
10	88·5–93·5	91	8	8 281	728	66 248
11	93·5–98·5	96	3	9 216	288	27 648
		Totals:	50		3 790	296 830

From the above table we have the totals $\sum_{i=1}^{11} x_i' f_i = 3790$ and $\sum_{i=1}^{11} x_i'^2 f_i = 296\,830$.

Substituting these values, along with $n = 50$, into Formula (4.19), we have

$$s^2 = \frac{i}{49}\left[296\,830 - \frac{(3790)^2}{50}\right].$$

$$= \frac{1}{49}[296\,830 - 287\,282]$$

$$= \frac{1}{49}[9548]$$

$$s^2 = 194\cdot9$$

$$s = \sqrt{194\cdot9} = 13\cdot96$$

The computations above were done by hand. They are not particularly lengthy, although a table of the squares of numbers will save a little time. When a desk calculator is used, the two sums $\sum_{i=1}^{k} x_i' f_i$ and $\sum_{i=1}^{k} x_i'^2 f_i$ may be found separately by cumulative multiplication.

When the data are unclassified the two necessary sums $\sum_{i=1}^{n} x_i$ and $\sum_{i=1}^{n} x_i^2$ may be found simultaneously on a desk calculator by cumulative squaring. The computations for s^2 for unclassified data, using the 50 sample observations given in Table 4.7, are shown below. The reader may verify them if he wishes to test his skill in using a desk calculator. We have

$$s^2 = \frac{1}{n-1}\left[\sum_{i=1}^{50} x_i^2 - \frac{\left(\sum_{i=1}^{50} x_i\right)^2}{n}\right]$$

$$= \frac{1}{49}\left[298\,168 - \frac{(3796)^2}{50}\right]$$

$$= \frac{1}{49}[298\,168 - 288\,192]$$

$$= \frac{1}{49}[9976]$$

$$s^2 = 203\cdot6$$

and $$s = \sqrt{203\cdot6} = 14\cdot7$$

Note that, although we are using the same sample data, the value of s^2 for classified data differs from the value of s^2 for unclassified data.

EXERCISES

1. Find the range, mean absolute deviation, variance, and standard deviation for each of the following samples:

(a) 5, 10, 11, 6, 13, 10, 8

(b) 31, 42, 37, 55, 70, 52

(c) 1·672, 1·541, 1·603, 1·659, 1·499, 1·591, 1·630

(d) 101·4, 150·1, 134·8, 139·2, 124·3

(e) 5, 8, 11, 9, 8, 6, 8

(f) 37, 35, 40, 35, 33, 36, 35

2. Find the variance of the following data, classfying the data if convenient:

395	369	374	348	373
376	348	360	386	377
372	337	378	359	351
367	376	380	368	382

3. Find the variance of the data that were classified in Exercise 8, Chapter III.

4. Find the variance and the standard deviation of the data that were classified in Exercise 10 of Chapter II.

CHAPTER V

ELEMENTARY PROBABILITY AND THE BINOMIAL DISTRIBUTION

INTRODUCTION

The idea of probability is a familiar one to everyone. Statements such as the following are heard frequently: 'You had better take an umbrella because it is likely to rain.' 'It is impossible to drive from here to London in three hours.' 'It is not likely to snow today.' 'I am almost certain that we will go home for the holidays.' 'The horse was 4 to 1 to win.' 'He will probably read at least three books during the next two weeks.' 'There will probably be some embarrassment if both of them attend the party.' 'The probability that we will go to war because of a nuclear accident is very low.'

The statements in the previous paragraph, although illustrating that most of us use the concept of probability in our everyday speech, also illustrate that there is a great deal of imprecision involved in such statements. For instance, even though the professional odds-makers assert that 'the odds that the horse will win the race are 4 to 1', everyone might not agree that the odds are, in fact, 4 to 1. Some might think that the odds are even; others, betting on the horse, perhaps might feel that the odds are at least 10 to 5; some of those betting against the horse might feel that another horse is really more likely to win. Or, suppose that a family has been arguing about whether they will go to the wife's home for the holidays, and have not reached a decision. Even though the wife may be able to say, 'It is almost certain that we will go home for the holidays,' the husband might not even think that such an occurrence is likely.

This sort of imprecision is intolerable in mathematics. Part of the imprecision in the above statements is due to the fact that the probability of a *unique* event is being discussed, and part of it comes from discussing an event for which there is more than one point of view. These difficulties can be avoided if we restrict our discussion of probability to events which are outcomes of experiments that can be repeated, and if we deal with idealized situations.

Some of the most convenient ideal situations are furnished by games of chance. It is for this reason that any discussion of probability seems to concern itself largely with cards and dice and drawing balls from urns, and not because statisticians are habitual gamblers. After the rudiments of probability are grasped from discussions of ideal situations we will attempt to transfer this knowledge to problems and experiments in the real world, where knowledge must be applied if it is to be of any practical value.

The terms 'experiment', 'outcome', and 'event' will not be defined in a formal way, although the following remarks may be helpful in understanding what these terms mean. We might think of an experiment as our being dealt a hand of five cards from the top of a well-shuffled pack of ordinary playing cards. If we consider each different set of five cards as a different outcome, then, using techniques which will be discussed later in this chapter, we can determine that there are 2 598 960 outcomes. Some events are that a hand will contain 2 aces and 2 non-aces, that a hand will contain 4 aces and a king, that a hand will contain no black cards, and so on. A certain number of outcomes correspond to each of these events. For instance, 4512 of the 2 598 960 possible hands contain 3 aces and 2 non-aces, and thus correspond to the event 'a hand will contain 3 aces and 2 non-aces'. When the experiment of dealing a five-card hand results in one of these 4512 hands we say that the event of obtaining 3 aces and 2 non-aces has occurred.

We can consider the toss of a pair of dice to be an experiment. We will see later that this experiment has 36 outcomes. Several events are throwing a total of 7, throwing an even total, throwing a total less than 6, and so on.

The material in this chapter builds from the probabilities of simple events, through the probabilities of combinations of events, to the binomial distribution, a very important theoretical distribution that is applicable to many real-life situations. Permutations and combinations are also discussed, because they are necessary for a complete development of the binomial distribution.

A knowledge of elementary probability is necessary to understand what the two types of error are when a hypothesis is tested, and what the critical region of a test is: it is extremely important that the terms 'Type I error', 'Type II error', and 'critical region' (all explained in Chapter VII) are understood.

Chapter VI, as well as the present chapter, deals with probability. There, the normal distribution and the normal approximation to the binomial are discussed. Therefore, the material treated in this chapter is used throughout Chapter VI.

The ideas and notation of this chapter are used in the derivation of confidence intervals for various parameters (a term that is defined later) in Chapter X. The idea of independence is referred to in the discussion of contingency tables in Chapter XI. The binomial distribution is used in the development of a test about the median of a population, known as the sign test, also in Chapter XI.

In addition to the fact that a background of knowledge about probability is necessary for understanding much of the later material in this book, the reader will probably encounter problems of his own in which as much probability as is discussed here is needed.

PROBABILITIES OF SIMPLE EVENTS

When no two outcomes of an experiment can occur at the same time the outcomes are said to be **mutually exclusive**. If no outcome is any more likely to occur than any other the outcomes are said to be **equally likely**. The usual way of determining whether the outcomes of an experiment are mutually exclusive and equally likely is by reason alone. One simply thinks about the experiment prior to performing it, and decides, on the basis of his reason alone, whether the outcomes are mutually exclusive and equally likely. Since this reasoning takes place prior to the performance of the experiment, it is called *a priori* reasoning.

We now state what is known as the classical definition of probability.

If an experiment can result in n equally likely, mutually exclusive outcomes, r of which correspond to the occurrence of some event A, then **the probability** that event A will occur, denoted $P(A)$, will be defined as the ratio r/n; symbolically,

$$P(A) = r/n \qquad (5.1)$$

The author recommends that you learn to express Formula (5.1) verbally; perhaps, 'The probability of Event A equals the number of outcomes favourable to A divided by the total number of outcomes.' Using this idea, but stating it much more briefly, it is helpful to think of Formula (5.1) in the form

$$P(A) = \frac{\text{Favourable outcomes}}{\text{Total outcomes}} \qquad (5.2)$$

Note that in order to use the above definition we must be dealing with equally likely, mutually exclusive events. Otherwise the definition is not valid. The only way one can determine whether the outcomes of an experiment are equally likely and mutually exclusive is by reasoning through the situation *a priori*. The process is not at all difficult, and in the more simple cases appears to be common sense.

For instance, consider an ordinary pack of playing cards. If our experiment consists of drawing one card at random from the well-shuffled pack, then there are 52 equally likely outcomes. There are 52 outcomes because the drawing might result in any one of the 52 cards. We decide that the outcomes are equally likely by reasoning that if the pack is well shuffled and a card is drawn at random there is certainly no reason to expect any card to be more likely to be drawn than any other. On any draw of a single card we get just one card and no other, so we conclude that the outcomes are mutually exclusive. Note that we decide that the probabilities are equally likely before the experiment—in fact, without even performing the experiment. This is *a priori* reasoning.

Of course, we could take an empirical approach and perform the experiment a great many times (1 000 000, say) and, upon noting that the proportion of times each card is drawn is approximately the same as that for every other card, conclude that any card is equally likely to be drawn. In fact, there is another approach to formulating a definition of the probability of Event A which is valid when the outcomes are not equally likely. The definition states that, as an experiment is performed more and more times, if the ratio of the number of occurrences of A to the total number of occurrences approaches some number, say, p, $0 \leq p \leq 1$, then p is called the probability of Event A.

In the following examples, and henceforth throughout the book, the probability that Event A will occur is denoted by the symbols $P(A)$. For instance, the probability of a toss of a coin resulting in heads could be denoted P (heads) or $P(H)$. What to write within the parentheses is determined by the individual. The notation is a shorthand, and can be made as brief as is desired so long as it is understandable to anyone who might read it.

EXAMPLE: Find the probability of drawing a black card in a single random draw from a well-shuffled pack of ordinary playing cards.

The 'experiment' is the drawing of a card. There are 52 cards; hence the experiment has 52 outcomes. We reason that each of these 52 outcomes is equally likely. Only one card can be drawn on any draw, so the outcomes are mutually exclusive. The 'event' in which we are interested is that of drawing a black card. There are 26 black cards in the pack, so the number of outcomes favourable to the occurrence of the event is 26. Thus, we have

$$P \text{ (drawing a black card)} = \frac{\text{Favourable outcomes}}{\text{Total outcomes}} = \frac{26}{52} = \frac{1}{2}$$

EXAMPLE: Find the probability of drawing a face card in a single random draw from a well-shuffled pack of ordinary playing cards.

The 'experiment' is the drawing of a card, and once again there are 52 mutually exclusive, equally likely outcomes. The 'event' in which we are interested is that of drawing a face card. There are three face cards in each suit—the jack, the queen, and the king. There are four suits, and hence twelve face cards in the pack. The probability that we seek is

$$P \text{ (drawing a face card)} = \frac{\text{Favourable outcomes}}{\text{Total outcomes}} = \frac{12}{52} = \frac{3}{13}$$

The next three examples also involve playing cards and can be discussed somewhat more briefly than the two preceding ones.

EXAMPLE: Find the probability of drawing a spade on a single random draw from a well-shuffled pack of cards.

The number of outcomes of the experiment is 52, of which 13 are favourable to the event of drawing a spade since there are 13 spades in the pack. Thus,

$$P \text{ (drawing a spade)} = \frac{\text{Favourable outcomes}}{\text{Total outcomes}} = \frac{13}{52} = \frac{1}{4}$$

EXAMPLE: Find the probability of drawing an ace from a pack of cards on a single draw.

There are four aces in the pack. Hence 4 of the 52 possible outcomes of the experiment correspond to the event of drawing an ace. Therefore, we have

$$P \text{ (drawing an ace)} = \frac{\text{Favourable outcomes}}{\text{Total outcomes}} = \frac{4}{52} = \frac{1}{13}$$

EXAMPLE: Find the probability of drawing the ace of spades from a pack of cards on a single draw.

There is only one ace of spades in the pack, and therefore only one of 52 possible outcomes is favourable to the event. We have

$$P \text{ (drawing the ace of spades)} = \frac{\text{Favourable outcomes}}{\text{Total outcomes}} = \frac{1}{52}$$

There are numerous other 'ideal' situations—involving dice, coins, balls of different colours, numbered tickets, and so on—which can be used to illustrate the finding of the probabilities of elementary events by application of the classical definition of probability. We will consider only one other example, involving numbered balls, although many of the other situations will be encountered later.

EXAMPLE: There are 15 balls, numbered from 1 to 15, in a bag. If a person selects one at random, what is the probability that the number printed on the ball will be a prime number greater than 5?

You may recall that a prime number is defined as a number greater than 1 which is divisible only by itself and 1. For example, 3, 5, 13, and 23 are prime numbers, whereas 9, 15, 21, and 27 are not.

The 'experiment' is the drawing of a ball, and it has 15 outcomes. The 'event' in which we are interested is that of drawing a ball that has a prime number greater than 5 printed on it. There are three favourable outcomes—drawing a ball numbered either 7, 11, or 13. Hence, we have

$$P \text{ (drawing a prime number} > 5) = \frac{\text{Favourable outcomes}}{\text{Total outcomes}} = \frac{3}{15} = \frac{1}{5}$$

PROBABILITIES OF TWO EVENTS

The examples in the preceding section illustrate the procedure for finding the probability of the occurrence of a simple event. To calculate the probabilities of more complicated events (which can be considered as combinations of simple events) we will need some rules. These rules will be stated without justification in the following paragraphs, although they are proved in *Mathematics Made Simple*, Chapter 18.

The 'probability of the occurrence of Event B, given that Event A has already occurred (or is certain to occur)' is called the **conditional probability** of B and is denoted $P(B|A)$. Except that the prior occurrence of Event A is taken into account, the probability that Event B will occur is determined exactly as in the preceding section.

As an example, suppose that an experiment consists of drawing two cards in succession from an ordinary pack of cards, without replacing the first card. Drawing in this manner is often called merely 'drawing without replacement'. Let Event A be the event of drawing a spade on the first draw, and Event B be the event that a spade is drawn on the second draw. If a spade is obtained on the first draw, then Event A has occurred. The drawn card is not replaced and only 51 cards remain, of which 12 are spades. The experiment of drawing a card from a pack of 51 cards has 51 equally likely, mutually exclusive outcomes, 12 of which are favourable to the event of drawing a spade. Thus, using the classical definition of probability, after the first draw has resulted in a spade the probability that the second draw will result in a spade is $\frac{12}{51}$. Symbolically, for the two events in the present discussion, $P(B|A) = \frac{12}{51}$.

The probability that *both* of the events A and B will occur (also called the **joint occurrence of Events A and B**) is denoted by $P(AB)$. Often it is helpful to think of the events as occurring in sequence—as A followed by B, say—even though such an order in time may not be implied in the problem. It can be shown that, for any two events A and B,

$$P(AB) = P(A) \cdot P(B|A) \tag{5.3}$$

Naturally it makes no difference which of the two events is called A and which is called B. It is helpful to become familiar with the substance of Formula (5.3) in words: 'The probability that Events A and B will both occur equals the probability that Event A will occur multiplied by the probability, calculated on the assumption that Event A has already occurred, that Event B will occur.'

EXAMPLE: Find the probability of drawing a spade on each of two consecutive draws from a standard pack, without replacement of the first card.
Define the events A and B as follows:

Event A: spade on first draw,
Event B: spade on second draw.

From Formula (5.3) we have $P(AB) = P(A) \cdot P(B|A)$. We must find $P(A)$ and $P(B|A)$.

The first draw can have any one of 52 outcomes, of which 13 are favourable to the event of drawing a spade. Then

$$P(A) = \tfrac{13}{52} = \tfrac{1}{4}$$

Given that a spade occurred on the first draw, the second draw has 51 outcomes, 12 of which are favourable to the event of drawing a spade. Hence,

$$P(B|A) = \tfrac{12}{51} = \tfrac{4}{17}$$

We have the probability that two consecutive spades are drawn is

$$P(AB) = P(A) \cdot P(B|A)$$
$$= \tfrac{1}{4} \cdot \tfrac{4}{17}$$
$$= \tfrac{1}{17}$$

EXAMPLE: Find the probability that a face card is drawn on the first draw and an ace on the second in two consecutive draws, without replacement, from a standard pack of cards.

Let the events A and B be defined as follows:

Event A: face card on first draw,
Event B: ace on second draw.

The first draw has 52 outcomes, and 12 of them are favourable to the event of drawing a face card. Thus,

$$P(A) = \tfrac{12}{52} = \tfrac{3}{13}$$

The second draw has 51 outcomes, of which 4 are favourable to the event of drawing an ace, and we have

$$P(B|A) = \tfrac{4}{51}$$

Hence, from Formula (5.3), the probability that we draw a face card and an ace, in that order, is

$$P(AB) = P(A) \cdot P(B|A)$$
$$= \tfrac{3}{13} \cdot \tfrac{4}{51}$$
$$= \tfrac{4}{221}$$

If the occurrence or non-occurrence of Event A has no effect on the occurrence of Event B, then Events A and B are said to be **independent**. When Events A and B are independent the probability of Event B is the same whether or not we are given any information about the occurrence or non-occurrence of Event A. Symbolically, if Events A and B are independent, then

$$P(A|B) = P(A) \tag{5.4}$$
and
$$P(B|A) = P(B)$$

Then, if Events A and B are independent (5.3) becomes

$$P(AB) = P(A) \cdot P(B) \tag{5.5}$$

If two events are not independent, then they are **dependent**. Although the independence of Events A and B can be ascertained by noting whether the probability of the occurrence of both events equals the product of the probabilities of the separate events, it is possible in many simple probability problems to determine *a priori* whether the events are independent. For example, suppose that Event A is drawing a spade in one draw from a pack of cards and that Event B is rolling a total of 7 with a pair of dice. Obviously the card drawn doesn't affect in any way the total spots appearing on a throw of two dice. Then we can say *a priori* that these two events are independent. On the other hand, if Event *A* is drawing an ace on the first draw from a pack of cards and Event *B* is drawing an ace on the second draw it is obvious that the occurrence or non-occurrence of the first event *does* affect the probability of the second. We can say *a priori* that these last two events are not independent. They are dependent events.

As an example of calculating the probability of the joint occurrence of two independent events, we will consider the events *A* and *B* of the first illustration in the preceding paragraph. First, however, it is necessary to discuss how to find the probabilities of obtaining the various totals when throwing a pair of dice.

A pair of dice is a pair of cubes, each of which has 1, 2, 3, 4, 5, and 6 spots on its six respective faces. On a throw of 2 dice, there are 11 possible totals of the numbers of spots showing on the cubes, from 2 through 12.

However, we cannot simply apply the classical definition of probability and say that the probability of throwing any specified total is $\frac{1}{11}$ because the 11 possible totals are not equally likely. In order to deal with outcomes which are equally likely we can consider the dice as being distinguishable—one being red, say, and the other, black—and think of them as being tossed one at a time rather than simultaneously. If the red die [*Die* is the singular form of the plural *dice* and refers to only one of the familiar cubes with the spots on each face] is tossed first there are six possible outcomes—either 1, 2, 3, 4, 5, or 6 will turn up. Suppose that 1 turns up when the red die is tossed; the black die can turn up either 1, 2, 3, 4, 5, or 6; this gives six different possible outcomes. If 2 is rolled with the red die, then either 1, 2, 3, 4, 5, or 6 can be rolled with the black die; this gives six more outcomes.

Since there are 6 possible outcomes of the black die for each of the 6 possible outcomes of the red die, there are 36 outcomes altogether. There is no reason to consider any of these outcomes any more likely than any other; hence, we conclude that the 36 outcomes are equally likely.

The total number of spots showing for each of the 36 outcomes is shown in Table 5.1.

Table 5.1. Total Showing on Two Dice

		Red die					
		1	2	3	4	5	6
	1	2	3	4	5	6	7
	2	3	4	5	6	7	8
Black	3	4	5	6	7	8	9
die	4	5	6	7	8	9	10
	5	6	7	8	9	10	11
	6	7	8	9	10	11	12

EXAMPLE: Find the probability that Event *A*, drawing a spade on a single draw from a pack of cards, and Event *B*, rolling a total of 7 on a single roll of a pair of dice, will both occur.

We have previously found the probability of drawing a spade to be $\frac{1}{4}$,

$$P(A) = \tfrac{1}{4}$$

Examining Table 5.1, we see that of the 36 outcomes possible when a pair of dice is rolled, six correspond to rolling a total of 7. (These outcomes occur in the diagonal which runs from lower left to upper right in the table.) Since each outcome is equally likely, we can use the classical definition of probability and say that the probability of throwing a total of 7 is $\frac{6}{36}$. Symbolically,

$$P(B) = \frac{6}{36} = \frac{1}{6}$$

We have previously reasoned that these two events are independent. Therefore, using Formula (5.5), we have

$$P(AB) = P(A) \cdot P(B)$$
$$= \frac{1}{4} \cdot \frac{1}{6}$$
$$= \frac{1}{24}$$

and we have shown that the probability of the joint occurrence of Events A and B is $\frac{1}{24}$.

EXAMPLE: Find the probability of throwing two consecutive totals of 7 — in two throws of the dice.
Define Events A and B as follows:

Event A: obtaining a total of 7 on the first throw.

Event B: obtaining a total of 7 on the second throw.

Assuming that an honest pair of dice is rolled in an unbiased manner (that is, no combination of spots is favoured over any other due to the way the dice are held, shaken, rolled from the hand, etc.), the outcome of any roll has no effect on the outcome of any other roll. Therefore, we conclude that Events A and B are independent. As in the previous example, the probability of rolling a total of 7 is $\frac{1}{6}$, and we have

$$P(A) = \frac{1}{6} \quad \text{and} \quad P(B) = \frac{1}{6}$$

Applying Formula (5.5), we have

$$P(AB) = P(A) \cdot P(B)$$
$$= \left(\frac{1}{6}\right)\left(\frac{1}{6}\right)$$
$$= \frac{1}{36}$$

The probability that *at least one* of the events A and B occurs is denoted by $P(A + B)$. For any two events A and B this probability is given by

$$P(A + B) = P(A) + P(B) - P(AB) \qquad (5.6)$$

Note that there are no restrictions on the events A and B; Formula (5.6) is valid for any two events, independent or dependent, mutually exclusive or not. Because A and B are not always independent, $P(AB)$ must be calculated from Formula (5.3) rather than from (5.5).

EXAMPLE: Find the probability that on a single draw from a pack of playing cards we draw a spade or a face card or both.
Define Events A and B as follows:

Event A: drawing a spade,

Event B: drawing a face card.

Saying that we draw a spade or a face card or both is equivalent to saying that at least one of the Events A and B occurs. The probability of the occurrence of at least one of the events A and B is given by Formula (5.6) above. First, we need to find $P(AB)$. From (5.3), $P(AB) = P(A) \cdot P(B|A)$. We have previously seen that the probability of drawing a spade is $\frac{13}{52}$.

$$P(A) = \tfrac{13}{52}$$

There are three face cards in each suit; therefore,

$$P(B|A) = \tfrac{3}{13}$$

Thus we have $\qquad\qquad P(AB) = \tfrac{13}{52} \cdot \tfrac{3}{13} = \tfrac{3}{52}$

Incidentally, in this particular problem and in many other problems the probability of the joint occurrence of two (or more) events can be found by simply enumerating (either mentally or using pencil and paper) the outcomes which correspond to the joint occurrence of both events. In this example the joint occurrence of the events 'drawing a spade' and 'drawing a face card' would be the event 'drawing a spade face card'. There are 3 spade face cards, so the probability of drawing one of them on a single draw from a pack of playing cards would be $\frac{3}{52}$, the same result as we obtained previously.

In order to find $P(A + B)$ we need to find three quantities: $P(A)$, the probability of Event B without considering Event B at all; $P(B)$, the probability of Event B without considering Event A; and $P(AB)$, the joint occurrence of the events A and B. We have already found $P(A)$ and $P(AB)$, and it remains to find $P(B)$. The probability of drawing a face card is $\frac{12}{52}$, as we have determined in previous examples:

$$P(B) = \tfrac{12}{52}$$

Therefore, we have as the probability that we will draw a spade, a face card, or both,

$$P(A + B) = P(A) + P(B) - P(AB)$$
$$= \tfrac{13}{52} + \tfrac{12}{52} - \tfrac{3}{52} = \tfrac{22}{52}$$
$$= \tfrac{11}{26}$$

Alternatively, there are 13 spade cards and 9 other face cards, so that there are 22 ways in which the event may occur. Therefore the probability of occurrence is $\frac{22}{52} = \frac{11}{26}$ as before.

EXAMPLE: What is the probability that on a throw of two dice at least one of the following two events occurs:

> Event A: an even total is thrown,
> Event B: a total larger than 8 is thrown.

Once again, we will first find the probability of the joint occurrence of Events A and B. There are 18 outcomes that yield an even total. Therefore, the probability that an even total is thrown is $\frac{18}{36}$:

$$P(A) = \tfrac{18}{36} = \tfrac{1}{2}$$

There are four outcomes that yield even totals exceeding 8 (the only even totals exceeding 8 are 10 and 12; there are three outcomes that yield a total of 10 and one outcome that yields a total of 12), so given that an even total is thrown, the probability that the total exceeds 8 is $\frac{4}{18}$:

$$P(B|A) = \tfrac{4}{18} = \tfrac{2}{9}$$

Then the probability of the joint occurrence of Events A and B is given by

$$P(AB) = P(A) \cdot P(B|A)$$
$$= \tfrac{1}{2} \cdot \tfrac{2}{9}$$
$$= \tfrac{1}{9}$$

We now have $P(A)$ and $P(B|A)$, and we must find $P(B)$, the probability that a total exceeding 8 is thrown, without considering whether the total is even. We find, upon referring to Table 5.1, that there are 10 outcomes that yield a total greater than 8. Therefore,

$$P(B) = \tfrac{10}{36} = \tfrac{5}{18}$$

Then the probability that on a single toss of a pair of dice an even total is thrown, a total larger than 8 is thrown, or both, is given by

$$P(A + B) = P(A) + P(B) - P(AB)$$
$$= \tfrac{1}{2} + \tfrac{5}{18} - \tfrac{1}{9} = \tfrac{12}{18}$$
$$= \tfrac{2}{3}$$

As before, an alternative approach is simply to count the outcomes corresponding to the occurrence of at least one of the events, if it is convenient to do so. In the present example if we count the outcomes for which the total is even, is greater than 8, or both (counting those outcomes for which both statements are true only one time), we find that the number of outcomes is 24, i.e. 2; 4, 4, 4; 6, 6, 6, 6, 6; 8, 8, 8, 8, 8; 9, 9, 9, 9; 10, 10, 10; 11, 11; 12, which immediately yields the required probability $\tfrac{24}{36} = \tfrac{2}{3}$, the same as by the other method.

Sometimes the occurrence of one event will automatically exclude the occurrence of another. For example, on a single draw from a pack of cards either Event A, a spade is drawn, or Event B, a heart is drawn, occurs; the occurrence of either event excludes the occurrence of the other. If two dice are rolled either Event A, a total of 5 is thrown, or Event B, a total other than 5 is thrown, occurs; here again, the occurrence of either event excludes the occurrence of the other. Such events are called **mutually exclusive events**. In other words, two events are called mutually exclusive events if they both cannot happen—either one or the other can occur, but not both.

Note that we are not guaranteed that either event will occur. In the first illustration in the preceding paragraph it is not certain that a draw will result in either a heart or a spade; a club or diamond could be drawn as well. If two (or more) events are defined so that at least one of them must occur when an experiment is performed they are called **exhaustive events**. If they are mutually exclusive as well as exhaustive they are called **mutually exclusive, exhaustive events**. The events of the second illustration in the preceding paragraph are mutually exclusive, exhaustive events. The total rolled on a single roll of a pair of dice must be either 5 or a total other than 5—one of these two events *must* occur. Therefore the events are exhaustive. We noted in the preceding paragraph that the events are mutually exclusive.

Event AB, the joint occurrence of Events A and B, is certain not to occur if Events A and B are mutually exclusive events. Whenever an event is impossible (certain not to occur) the probability of its occurrence is 0. (When an event is certain not to occur there are no outcomes favourable to that event, and by the classical definition of probability the probability of that event is $0/n = 0$, where n is the total number of outcomes.) Symbolically,

$$P(AB) = 0$$

and substituting this value for $P(AB)$ in Formula (5.6), we have

$$P(A + B) = P(A) + P(B) \qquad (5.7)$$

when Events A and B are mutually exclusive.

EXAMPLE: What is the probability that on a single draw at random from a pack of cards either a black face card or a red ace will be drawn?

The two events 'drawing a black face card' and 'drawing a red ace' cannot both occur, and are therefore mutually exclusive. We have

$$P \text{ (black face card)} = P(A) = \tfrac{6}{52}$$
$$P \text{ (red ace)} = P(B) = \tfrac{2}{52}$$

Applying Formula (5.7), we have

$$P(\text{black face card or red ace}) = P(A+B) = P(A) + P(B)$$
$$= \tfrac{6}{52} + \tfrac{2}{52} = \tfrac{8}{52}$$
$$= \tfrac{2}{13}$$

Note that the two events A and B are not exhaustive.

Alternatively, there are 6 black face cards and 2 red aces, so that the event may occur in 8 ways out of 52. Therefore the probability of the occurrence is $\tfrac{8}{52} = \tfrac{2}{13}$ as before.

PROBABILITIES FOR COMBINATIONS OF THREE OR MORE EVENTS

The formulae stated in the preceding section, which give the probability of the joint occurrence of two events A and B and the probability of the occurrence of at least one of the events A and B, generalize to three or more events.

The probability that each of the events A, B, and C occurs is denoted by $P(ABC)$ and is given by

$$P(ABC) = P(A) \cdot P(B|A) \cdot P(C|AB) \tag{5.8}$$

where $P(C|AB)$ denotes the probability that Event C occurs given that Events A and B have already occurred.

EXAMPLE: Find the probability that three successive face cards are drawn in three successive draws (without replacement) from a pack of cards.

Define Events A, B, and C as follows:

   Event A: a face card is drawn on the first draw,

   Event B: a face card is drawn on the second draw,

   Event C: a face card is drawn on the third draw.

Recalling that twelve of the 52 cards are face cards, we have

$$P(A) = \tfrac{12}{52}$$

The probability that the second draw results in a face card, given that the first draw resulted in a face card, is $\tfrac{11}{51}$ (only 11 of the remaining cards are face cards) and we have

$$P(B|A) = \tfrac{11}{51}$$

Given that the first two draws have resulted in face cards, the probability that the third draw results in a face card is $\tfrac{10}{50}$ (only 10 of the remaining cards are face cards) and we have

$$P(C|AB) = \tfrac{10}{50}$$

Substituting these values into Formula (5.8), we have the probability of the joint occurrence of Events A, B, and C,

$$P(ABC) = P(A) \cdot P(B|A) \cdot P(C|AB)$$
$$= \tfrac{12}{52} \cdot \tfrac{11}{51} \cdot \tfrac{10}{50}$$
$$= \tfrac{11}{1105}$$
$$= 0 \cdot 010$$

If we performed the experiment of drawing three cards at random from a well-shuffled pack a great many times we would expect to draw three consecutive face cards only about one time in a hundred, on the average.

When three events A, B, and C are independent the outcome of any event or pair of events does not affect the outcome of any other event. Then we have

$$P(B|A) = P(B)$$

and $$P(C|AB) = P(C)$$

and Formula (5.8) becomes

$$P(ABC) = P(A) . P(B) . P(C) \qquad (5.9)$$

EXAMPLE: Find the probability of drawing three consecutive face cards on three consecutive draws (with replacment) from a pack of cards.

Since after each draw the card is replaced and the pack reshuffled before the next card is drawn, what is drawn on any draw has no effect on what is drawn on any other draw. Therefore if we define

Event A: face card on first draw,

Event B: face card on second draw, and

Event C: face card on third draw,

we have $$P(ABC) = P(A) . P(B) . P(C)$$
$$= \tfrac{12}{52} . \tfrac{12}{52} . \tfrac{12}{52}$$
$$= \tfrac{27}{2197} = 0.012$$

We see that when the cards are replaced we are more likely to draw three consecutive face cards than we are when the cards are not replaced.

The probability that at least one of the events A, B, and C will occur is not as straightforward a generalization of the two-event formula as is the above formula for the probability that *each* of the events A, B, and C will occur. It can be shown that the probability that at least one of the events A, B, and C will occur is given by

$$P(A + B + C)$$
$$= P(A) + P(B) + P(C) - P(AB) - P(AC) - P(BC) + P(ABC) \quad (5.10)$$

The analogous expression for $P(A + B + C + D)$ has 15 terms, and that for $P(A + B + C + D + E)$ has 31 terms. The reader will doubtless agree that the computation of probabilities by means of a formula such as (5.10) becomes tedious rather quickly. If the events are mutually exclusive, however, the required probability is very easy to compute, the formula being simply a straightforward extension of Formula (5.7). We would have

$$P(A + B + C) = P(A) + P(B) + P(C) \qquad (5.11)$$
$$P(A + B + C + D) = P(A) + P(B) + P(C) + P(D)$$

and so on, since the probabilities that any two or three of the events would occur jointly would all be zero.

EXAMPLE: A card is drawn at random from a pack of cards. Find the probability that at least one of the following three events will occur:

Event A: a heart is drawn.

Event B: a card which is not a face card is drawn.

Event C: the number of spots (if any) on the drawn card is divisible by 3.

We need the probabilities of seven events—$P(A)$, $P(B)$, $P(C)$, $P(AB)$, $P(AC)$, $P(BC)$, and $P(ABC)$.

There are thirteen of the 52 possible outcomes favourable to the event of drawing a heart, forty of the 52 outcomes favourable to the event of drawing a card that is not a face card, and twelve of the 52 outcomes favourable to the event of drawing a card that has a number of spots that is divisible by 3. Therefore we have

$$P(A) = \tfrac{13}{52}$$
$$P(B) = \tfrac{40}{52}$$
and $$P(C) = \tfrac{12}{52}$$

The reader may verify the other four probabilities, the calculations of which are shown below:

$$P(AB) = P(A) \cdot P(B|A) = \tfrac{13}{52} \cdot \tfrac{10}{13} = \tfrac{10}{52}$$
$$P(AC) = P(A) \cdot P(C|A) = \tfrac{13}{52} \cdot \tfrac{3}{13} = \tfrac{3}{52}$$
$$P(BC) = P(B) \cdot P(C|B) = \tfrac{40}{52} \cdot \tfrac{12}{40} = \tfrac{12}{52}$$
$$P(ABC) = P(A) \cdot P(B|A) \cdot P(C|AB)$$
$$= \tfrac{13}{52} \cdot \tfrac{10}{13} \cdot \tfrac{3}{10}$$
$$= \tfrac{3}{52}$$

Then, applying Formula (5.10), we have

$$P(A + B + C) = \tfrac{13}{52} + \tfrac{40}{52} + \tfrac{12}{52} - \tfrac{10}{52} - \tfrac{3}{52} - \tfrac{12}{52} + \tfrac{3}{52}$$
$$= \tfrac{43}{52}$$

Fortunately, there is another general approach that is usually easier. It is certain that any event, E, will either occur or will not occur. That is, the probability of the occurrence or the non-occurrence of E is equal to 1. Designating the non-occurrence of E (also called the event not-E) as $\bar{E}$, we have

$$P(E + \bar{E}) = 1 \qquad (5.12)$$

The events E and $\bar{E}$ are mutually exclusive, exhaustive events. Applying Formula (5.7), (5.12) becomes

$$P(E) + P(\bar{E}) = 1 \qquad (5.13)$$

Events E and $\bar{E}$ are called **complementary events**; the outcomes of one complement the outcomes of the other because together they exhaust all possible outcomes. Solving Formula (5.13) for $P(E)$, we obtain

$$P(E) = 1 - P(\bar{E}) \qquad (5.14)$$

In words, the probability of the occurrence of any event equals one minus the probability of the occurrence of its complementary event.

Now let us apply this newly found knowledge to the problem of finding the probability of at least one of the events A, B, and C, $P(A + B + C)$. Either at least one of the events A, B, and C occurs or none of the events A, B, and C occurs. Therefore the complementary event of the event 'at least one of the events A, B, and C occurs' is the event 'none of the events A, B, and C occurs'. The probability that none of the events A, B, and C occurs is equivalent to the probability that Event A does not occur and Event B does not occur and Event C does not occur, which is equivalent to saying the probability of the joint occurrence of Events not-A, not-B, and not-C, written symbolically as $P(\bar{A}\bar{B}\bar{C})$. Thus, since the events $A + B + C$ and $\bar{A}\bar{B}\bar{C}$ are complementary events, we can use Formula (5.14) and obtain

$$P(A + B + C) = 1 - P(\bar{A}\bar{B}\bar{C}) \qquad (5.15)$$

We know how to find $P(\bar{A}\bar{B}\bar{C})$. Applying Formula (5.8), we have

$$P(\bar{A}\bar{B}\bar{C}) = P(\bar{A}) \cdot P(\bar{B}|\bar{A}) \cdot P(\bar{C}|\bar{A}\bar{B}) \qquad (5.16)$$

where   Event $\bar{A}$ is drawing a club, diamond, or spade;
       Event $\bar{B}$ is drawing a face card; and
       Event $\bar{C}$ is drawing a card the number of whose spots (if any) is not divisible by 3.

Thus, 39 of the possible 52 outcomes are favourable to drawing a club, diamond, or spade, and we have

$$P(\bar{A}) = \tfrac{39}{52}$$

Given that the card drawn is a club, diamond, or spade, the probability that the card is also a face card is $\tfrac{9}{39}$ since nine of the 39 possible outcomes are face cards, and we have

$$P(\bar{B}|\bar{A}) = \tfrac{9}{39}$$

Given that the card drawn is a face card from the club, diamond, or spade suit, the probability that the number of spots (if any) is not divisible by 3 is 1, because there are no spots on face cards, and we have

$$P(\bar{C}|\bar{A}\bar{B}) = 1$$

Substituting the above values into Formula (5.16), we have

$$P(\bar{A}\bar{B}\bar{C}) = P(\bar{A}) \cdot P(\bar{B}|\bar{A}) \cdot P(\bar{C}|\bar{A}\bar{B})$$
$$= \tfrac{39}{52} \cdot \tfrac{9}{39} \cdot 1$$
$$= \tfrac{9}{52}$$

Therefore, Formula (5.15) becomes

$$P(A + B + C) = 1 - P(\bar{A}\bar{B}\bar{C})$$
$$= 1 - \tfrac{9}{52}$$
$$= \tfrac{43}{52} \quad .$$

which agrees with the answer we obtained by the straightforward approach.

The next example illustrates the calculation of the probability of at least one of the events A, B, and C, when the three events are mutually exclusive.

EXAMPLE: Find the probability of throwing at least one of the following totals on a single throw of a pair of dice: a total of 5, a total of 6, a total of 7.

Define the events A, B, and C as follows:

   Event A: a total of 5 is thrown,
   Event B: a total of 6 is thrown,
   Event C: a total of 7 is thrown.

These events are mutually exclusive because the occurrence of any one excludes the occurrence of any of the others. (They are not exhaustive events, however.) Referring to Table 5.1, we see

$$P(A) = \tfrac{4}{36}$$
$$P(B) = \tfrac{5}{36}$$

and
$$P(C) = \tfrac{6}{36}$$

Applying Formula (5.11), we find that the probability of at least one of the events A, B, and C is

$$P(A + B + C) = P(A) + P(B) + P(C)$$
$$= \tfrac{4}{36} + \tfrac{5}{36} + \tfrac{6}{36} = \tfrac{15}{36}$$
$$= \tfrac{5}{12}$$

Now try Questions 1–10 of the Exercises at the end of this chapter.

PERMUTATIONS

In order to find the probability of an event A, we must be able to find the total number of outcomes and the number of outcomes favourable to A. In the examples that we have discussed, these numbers were easy to find. However, in many problems there are a large number of outcomes, and it is not possible to count them as we have been doing. With the proper mathematical ideas and techniques at our disposal, however, finding the numbers of outcomes becomes a rather simple task. We simply recognize what category the problem fits into, and then apply the appropriate technique (i.e. formula) to find the number of outcomes. Most counting difficulties fit into one or the other of two categories: they are either **permutation** problems or **combination** problems. The following material will enable us to determine the numbers of outcomes when these numbers are not small.

A permutation of objects is simply an arrangement of those objects. The usual situation is that we have n distinct objects at our disposal and wish to determine how many arrangements are possible if each arrangement is to consist of r objects ($r \leq n$). Think of r empty positions in a straight line.

Fig. 5.1

We wish to fill these r positions, and in so doing count the number of ways in which they can be filled. This number is the number of permutations of n objects selected r at a time. As an aid, we will make use of the following principle, stated without proof.

Fundamental Principle. If a first operation has m outcomes, and for each of these m outcomes a second operation has n outcomes, then Operation 1 followed by Operation 2 has mn outcomes. This principle can be extended to three or more operations.

EXAMPLE: Suppose that, for a certain luncheon, a man has a choice of one of three appetizers (soup, orange juice, grapefruit), a choice of one of three entrees (fish, ham, beef), and a choice of one of two desserts (pudding, ice-cream). How many different lunches are possible?

Operation 1, choosing an appetizer, has 3 outcomes. For each of these outcomes Operation 2, choosing an entree, has 3 outcomes. So there are 9 possible ways to chooze an appetizer and an entree. For each of these 9 possibilities, there are 2 outcomes for Operation 3, choosing a dessert. Thus, there are $3 . 3 . 2 = 18$ different possible lunches by the fundamental principle.

Frequently, a diagram known as a **tree diagram** is used to determine the number of different outcomes, or to display the various outcomes. The tree diagram for the present example is shown in Fig. 5.2.

The diagram in Fig. 5.2 illustrates very clearly why it is called a tree diagram, and why the outcomes of the separate operations are multiplied together in the fundamental principle. Examining Fig. 5.2 we see that each outcome of Operation 1 is a large branch. There are as many smaller branches sprouting from each large branch as there are outcomes of Operation 2. There are as many twigs on each smaller branch as there are outcomes of Operation 3. In order to find the total number of outcomes, all we need do is to count the twigs. We verify that there are eighteen outcomes.

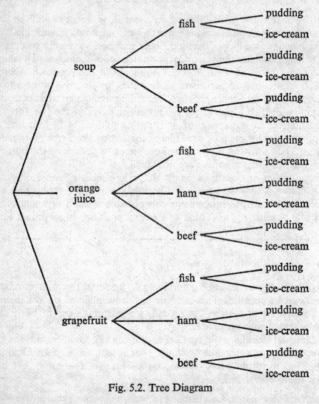

Fig. 5.2. Tree Diagram

In addition to enabling us to find the number of outcomes, a tree diagram can also be used as an aid in calculating the probabilities of combinations of simple events.

We can think of filling the r positions in Fig. 5.1 as performing r operations —Operation 1 is filling the first position, Operation 2 is filling the second position, and so on—and use the fundamental principle extended to r operations in order to calculate the number of ways these operations can be performed. That is, the number of ways r objects selected from n can be arranged equals the number of ways the r positions can be filled, which equals the number of ways Operation 1 can be performed multiplied by the number of ways Operation 2 can be performed, and so on.

We have n choices of an object to put into the first position, so Operation 1 has n outcomes. After the first position is filled, $n - 1$ objects remain, so the second position can be filled in $n - 1$ ways. The third position can be filled in $n - 2$ ways, the fourth in $n - 3$ ways, and so on. To find the number of ways in which the rth and last position can be filled we reason as follows: $r - 1$ positions have already been filled and hence $r - 1$ objects have been used. Then $n - (r - 1) = n - r + 1$ objects remain, and the rth position can be filled in $n - r + 1$ ways. Then the number of ways in which Operation 1 followed by Operation 2, ..., followed by Operation r can be performed is

$$n(n - 1)(n - 2) \ldots (n - r + 1)$$

(See *Mathematics Made Simple*, Chapter 17.)

To summarize the preceding discussion: The number of permutations of n distinct objects taken r at a time, denoted P_r^n, is given by

$$P_r^n = n(n - 1) \ldots (n - r + 1) \tag{5.17}$$

Instead of P_r^n, the symbols $_nP_r$ and $P(n,r)$ are frequently used to denote the number of permutations of n distinct objects taken r at a time. Different authors frequently use different symbols; the three given are all in common usage.

A formula for P_r^n that is neater in appearance than (5.17) can be derived if we multiply and divide the right-hand side by $(n - r)(n - r - 1) \ldots 2.1$. Thus we have

$$P_r^n = n(n - 1) \ldots (n - r + 1) \times \frac{(n - r)(n - r - 1) \ldots 2.1}{(n - r)(n - r - 1) \ldots 2.1}$$

or $\quad P_r^n = \dfrac{n(n - 1) \ldots (n - r + 1)(n - r)(n - r - 1) \ldots 2.1}{(n - r)(n - r - 1) \ldots 2.1} \tag{5.18}$

NOTE: The symbol $n!$ is read 'n factorial', and denotes the product of all the positive integers from 1 up to n if n is a positive integer. If n equals zero, $n!$ is defined to equal 1. The symbol $n!$ has no meaning if n represents anything other than a positive integer or zero. Symbolically, by definition

$$n! = 1.2.3 \ldots n \quad \text{if } n \text{ is a positive integer}$$
$$n! = 1 \quad \text{if } n = 0.$$

EXAMPLES:
$$4! = 1.2.3.4 = 24$$
$$7! = 1.2.3.4.5.6.7 = 5040$$
$$10! = 1.2.3 \ldots 10 = 3\ 628\ 800$$
$$0! = 1 \text{ (by definition)}$$
$$(\tfrac{1}{2})! = \text{meaningless (not defined)}$$
$$(-3)! = \text{meaningless (not defined)}$$
$$r! = 1.2.3 \ldots r, \text{ where } r \text{ is a positive integer}$$
$$(n - r)! = 1.2.3 \ldots (n - r), \text{ where } n - r \text{ is a positive integer}$$

Recalling the definition of $n!$ we see that the numerator of Formula (5.18) is simply $n!$ and the denominator is $(n - r)!$. Then we can write that formula more neatly as

$$P_r^n = \frac{n!}{(n - r)!} \tag{5.19}$$

This formula not only has a more pleasing appearance than Formula (5.17) but is easier to remember. We will use Formula (5.19) to derive an important formula in the next section, a formula that looks very much like Formula (5.19) and is easy to remember because of this fact. This furnishes yet another reason for deriving the formula for P^n_r that appears in (5.19), rather than being content with the formula in (5.17).

EXAMPLE: If there are 15 students in a certain class and there are 6 desks in the front row, how many different arrangements of students in the seats of the front row are possible? (Assume that no seats remain empty in the front row.)

The total number of 'objects' from which we are selecting is 15, and we are arranging groups of 6 'objects'; therefore, $n = 15$ and $r = 6$. Then, using Formula (5.9), the number of arrangements is

$$
\begin{aligned}
P^{15}_6 &= \frac{15!}{9!} \\
&= \frac{1 \cdot 2 \cdot 3 \cdot 4 \cdot 5 \cdot 6 \cdot 7 \cdot 8 \cdot 9 \cdot 10 \cdot 11 \cdot 12 \cdot 13 \cdot 14 \cdot 15}{1 \cdot 2 \cdot 3 \cdot 4 \cdot 5 \cdot 6 \cdot 7 \cdot 8 \cdot 9} \\
&= 10 \cdot 11 \cdot 12 \cdot 13 \cdot 14 \cdot 15 \\
&= 3\ 603\ 600
\end{aligned}
$$

Note that the number of permutations of n objects taken n at a time is simply $n!$. This can be seen most clearly, perhaps, from Formula (5.17) and the discussion immediately preceding it. When $r = n$, Formula (5.17) becomes

$$P^n_n = n(n - 1) \ldots 3 \cdot 2 \cdot 1 = n! \qquad (5.20)$$

EXAMPLE: If five persons are to pose for a photograph by standing in a row, how many different arrangements are possible?

Since we are arranging five 'objects' 5 at a time, the number of different arrangements is given by

$$
\begin{aligned}
P^5_5 &= 5! \\
&= 1 \cdot 2 \cdot 3 \cdot 4 \cdot 5 \\
&= 120
\end{aligned}
$$

There are also formulae for the number of permutations of n distinct objects arranged in a circle, and for the number of permutations of n objects, some of which are alike, but we will not treat these situations in this book.

COMBINATIONS

If order is of no importance, then we have a **combination** rather than a permutation. A **combination of** n **objects taken** r **at a time** is a selection of r objects taken from the n, without regard to the order in which they are selected or arranged. Order is irrelevant.

To emphasize the difference between a combination and a permutation, let us consider the following illustration.

Suppose that we have four objects, labelled A, B, C, and D. One selection of three objects out of the four (there are three other such selections) is the set

ABC

It makes no difference whether we write ABC, or BCA, or CAB—we still have the same combination of objects because order is unimportant. The only thing that matters is that object A is present, object B is present, and object C is present. However, using Formula (5.20), we see that the three letters A, B, and C can be arranged in $3! = 6$ ways. The six possible permutations or arrangements are ABC, ACB, BAC, BCA, CAB, CBA, i.e. 6 permutations but only 1 combination.

Each of the other three combinations—ABD, ACD, and BCD—also yields six permutations. The four possible combinations, along with the permutations that can be formed from each, are shown below.

ABC	ABD	ACD	BCD
ABC	ABD	ACD	BCD
ACB	ADB	ADC	BDC
CAB	DAB	DAC	DBC
CBA	DBA	DCA	DCB
BAC	BAD	CAD	CBD
BCA	BDA	CDA	CDB

We see that there are 24 permutations and that, in general, we can find the number of permutations by multiplying the number of combinations by the number of permutations contained in each combination. For the above illustration,

$$P_3^4 = 4 \cdot 6$$
$$= 24$$

This agrees with the result obtained by using Formula (5.19) (where $n = 4$, $r = 3$):

$$P_3^4 = \frac{4!}{1!}$$
$$= \frac{24}{1}$$
$$= 24$$

The approach used in the previous illustration furnishes an admirable method by which we can attack the general question: how many different combinations of r objects can be selected from a set of n objects?

We would like to find the number of sets of r objects that can be selected from n distinct objects, without regard to their order. Denote this number by the symbol C_r^n. Suppose that we actually select a set of r objects from the n—then we have *one* of the C_r^n possible combinations of r objects that can be selected from the n. After we have selected this particular set of r objects, suppose that we decide that order *is* important, and ask ourselves in how many ways we can arrange the r selected objects.

By Formula (5.20), r objects can be arranged in $P_r^r = r!$ ways. Therefore, our particular combination, and every other combination of r objects, yields $r!$ permutations of r objects.

The total number of combinations of r objects that can be selected from n is C_r^n. Each of these C_r^n combinations yields $r!$ permutations. Therefore, if we multiply each combination of r objects by the number of permutations of r

objects that it yields, then the resulting product gives the total number of permutations of r objects that can be selected from n objects. More briefly,

Total number of combinations ×
Number of permutations in **each** combination
= Total number of permutations   (5.21)

This procedure is exactly analogous to that of changing days to hours:

No. of days × hours in each day = No. of hours

As before, denoting the number of permutations of n objects taken r at a time by P_r^n and the number of combinations of n objects taken r at a time by C_r^n, Formula (5.21) becomes

$$(C_r^n) \cdot r! = P_r^n \qquad (5.22)$$

From Formula (5.19) we know that the number of permutations of n objects taken r at a time is given by

$$P_r^n = \frac{n!}{(n-r)!}$$

We now have P_r^n equal to two quantities. Equating these two equivalent quantities we have

$$(C_r^n) \cdot r! = \frac{n!}{(n-r)!}$$

which, when solved for C_r^n, yields

$$C_r^n = \frac{n!}{r!(n-r)!} \qquad (5.23)$$

Thus, we have derived a formula that gives the number of combinations. Note the similarity of Formula (5.23), which gives the number of combinations, with Formula (5.19), which gives the number of permutations.

Some other symbols which are used to denote the number of combinations of n objects taken r at a time are $_nC_r$, $C(n,r)$, and $\binom{n}{r}$. The last symbol will be used in the treatment of the binomial distribution later in this chapter.

EXAMPLE: If a club has a membership of ten, how many three-man committees are possible?

Order is not important, so this is a combination problem. The number of possible committees is equal to the number of ways three persons can be selected from ten persons, namely C_3^{10}. We have

$$C_3^{10} = \frac{10!}{3!7!}$$
$$= \frac{1.2.3.4.5.6.7.8.9.10}{1.2.3.1.2.3.4.5.6.7}$$
$$= \frac{8.9.10}{1.2.3}$$
$$= 120$$

Note that the seven factors of the 7! divide out the first seven factors of the 10!.

EXAMPLE: How many different bridge hands are there?

Bridge is played with a thirteen-card hand dealt from a standard pack of 52 cards. The order in which the cards are dealt is not important—we are interested only in which set of 13 cards we have. The number of different hands that might be *dealt* is equal to the number of different hands it is possible to *select*, if one were allowed to select his 13 cards. The number of 13-card hands that can be selected equals the number of combinations of 52 objects taken 13 at a time,

$$C^{52}_{13} = \frac{52!}{13!39!}$$

The factorials are so large in this problem that it is tedious to calculate this number directly as we did C^{10}_3 above. A table that gives the approximate values (and the logarithms) of $n!$, usually for values of n from 1 to 100, appears in most books of standard mathematical tables. Using such a table to calculate C^{52}_{13}, we find that the number of possible 13-card bridge hands is about 635 000 000 000.

Many problems involve both the use of the Fundamental Principle, stated near the beginning of the preceding section, and also the use of the formulae for permutations and/or combinations in order to 'count' large numbers of outcomes. The following two examples are representative of this general class of problems.

EXAMPLE: Of the approximately 635 000 000 000 different bridge hands, how many contain a 10-card suit?

As in the preceding example, since the number of ways one can be *dealt* such a hand is the same as the number of ways one could *select* such a hand, we will use the word 'select' in the following discussion.

We can think of the selection of a hand containing 10 cards in a single suit as being performed in three successive 'operations'. The selection of one of the four suits in which we wish to have 10 cards is Operation 1. From the previous discussion we know that the number of ways we can select one object from four objects is C^4_1. Therefore, Operation 1 has C^4_1 outcomes. Operation 2 is choosing 10 cards that we wish our hand to contain from the 13 cards of the selected suit. This can be done in C^{13}_{10} ways. Operation 3 is selecting the additional 3 cards that our hand must contain from the 39 cards of the 3 suits other than the suit from which we have selected 10 cards. This can be done in C^{39}_3 ways.

Then, by the Fundamental Principle, the number of different hands containing 10 cards in one suit equals the product of the number of outcomes of Operation 1 times the number of outcomes of Operation 2 times the number of outcomes of Operation 3. This product is

$$C^4_1 \cdot C^{13}_{10} \cdot C^{39}_3 = (4) \cdot (286) \cdot (9\,139)$$
$$= 10\,455\,016$$

Incidentally, the probability of being dealt a hand containing 10 cards only of the same suit is about 0·000 02.

EXAMPLE: A club that is composed of 50 married men, 45 married women, and 35 unmarried members plans to have a party. The party committee is to be composed of 3 married men, 2 married women, and 2 unmarried members. How many different committees are possible?

Selecting the proper number of committeemen from each class can be thought of as a sequence of three operations, and the number of different committees possible is equal to

$$C^{50}_3 \cdot C^{45}_2 \cdot C^{35}_2 = \frac{50 \cdot 49 \cdot 48}{3 \cdot 2 \cdot 1} \cdot \frac{45 \cdot 44}{2} \cdot \frac{35 \cdot 34}{2}$$
$$= 11\,545\,380\,000$$

MORE PROBABILITY

We are now in a position to calculate the probabilities of events that have a large number of outcomes. For example, suppose we want to find the probability that a hand of 13 cards dealt from a shuffled pack will contain 8 diamonds. As before,

$$P(\text{Event A}) = \frac{\text{Outcomes favourable to Event A}}{\text{Total outcomes}}$$

The total number of outcomes is equal to the number of ways 13 cards can be selected from 52 cards. We know that this number is the number of combinations of 52 things taken thirteen at a time—C_{13}^{52}.

To count the number of favourable outcomes is somewhat more difficult. We want to count the number of 13-card hands that are composed of 8 diamonds and 5 non-diamonds. This number is equivalent to the number of ways we can select a set of 13 cards composed of 8 diamonds and 5 non-diamonds. It is helpful to think of this selection as occurring in two stages— Operation 1, selecting the 8 diamonds, and Operation 2, selecting the 5 non-diamonds.

Operation 1 can be performed in C_8^{13} ways, and Operation 2 can be performed in C_5^{39} ways. For each of the C_8^{13} sets of diamonds there are C_5^{39} sets of non-diamonds. Hence the number of favourable outcomes is $(C_8^{13})(C_5^{39})$, and the probability that a hand of 13 cards will consist of 8 diamonds and 5 non-diamonds is

$$\frac{C_8^{13} \cdot C_5^{39}}{C_{13}^{52}}$$

which is approximately equal to 0·001.

THE BINOMIAL DISTRIBUTION

Consider the following problem: there are 3 red balls and 2 white balls in a box. An experiment consists of drawing 3 balls in succession, with replacement. What is the probability that no red balls will be obtained in the 3 draws? One red ball? Two red balls? Three red balls?

The probability that a red ball is drawn on the first draw is $\frac{3}{5}$. Since the drawn ball is replaced before another ball is drawn at random, the probability that the second (and also the third) draw results in a red ball is also $\frac{3}{5}$. The probability that a white ball is drawn on any draw is $\frac{2}{5}$. The probability of obtaining no red balls and the probability of obtaining 3 red balls can be found directly without too much trouble. However, we will follow a less direct approach in order to illustrate some general procedures.

A tree diagram can be used to display all possible outcomes of this experiment. Let R denote the event that a red ball has been drawn, and W, the event that a white ball has been drawn. See Fig. 5.3.

The probabilities for the outcomes of each particular draw are put on the appropriate branches of the tree. The experiment itself is made up of three operations (the three draws) and the total number of outcomes of the experiment can be found by counting the number of branches on the right-hand side of Fig. 5.3. Thus, we see that there are eight outcomes for this experiment. A convenient way to denote any particular outcome is by three letters representing the colours of the balls obtained on the three respective draws. Thus, *RWW* would denote the event of drawing a red ball on the first draw, a white ball on the second draw, and a white ball on the third draw.

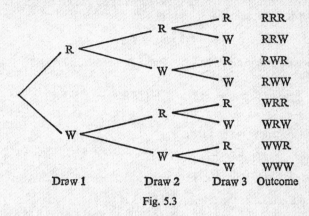

Draw 1   Draw 2   Draw 3   Outcome

Fig. 5.3

If we wish to find $P(RWW)$ we can use the formulae given in the section on Permutations. From Formula (5.8) we have

$$P(RWW) = P(R) \cdot P(W|R) \cdot P(W|RW)$$

These probabilities are already displayed on the branches of the tree diagram. In order to find the joint occurrence of any specified sequence of simple events, such as *RWW*, all we need do is to follow that sequence through the tree diagram, multiplying the probabilities written on the branches. To find $P(RWW)$ we multiply $\frac{3}{5}$ (first branch) by $\frac{2}{5}$ (second branch) by $\frac{2}{5}$ (third branch), obtaining

$$P(RWW) = \frac{3}{5} \cdot \frac{2}{5} \cdot \frac{2}{5} = \frac{12}{125}$$

Of course, $P(RWW)$ is not difficult to find, even without the aid of the tree diagram. One of the major advantages of the tree diagram is that the entire situation is conveniently displayed. Although the replacement of balls makes the probabilities the same from draw to draw, and the outcome of each draw is independent of those of the others, the tree diagram is applicable to

Table 5.2

Outcome	Probability
RRR	$\frac{27}{125}$
RRW	$\frac{18}{125}$
RWR	$\frac{18}{125}$
RWW	$\frac{12}{125}$
WRR	$\frac{18}{125}$
WRW	$\frac{12}{125}$
WWR	$\frac{12}{125}$
WWW	$\frac{8}{125}$

situations in which outcomes are not independent and in which probabilities change. The outcomes and their probabilities are shown in Table 5.2.

These outcomes are mutually exclusive and exhaustive. Therefore, to find the probability of any event, we simply add the probabilities associated with the outcomes that correspond to the occurrence of that event. Only one outcome, *RRR*, corresponds to the event of obtaining three red balls, and we have

$$P(3 \text{ red}) = \tfrac{27}{125}$$

Three outcomes correspond to obtaining two red balls, *RRW*, *RWR*, and *WRR*. Thus,

$$P(2 \text{ red}) = P(RRW) + P(RWR) + P(WRR)$$
$$= \tfrac{54}{125}$$

In a similar manner we find that

$$P(1 \text{ red}) = \tfrac{36}{125}$$

and that $$P(0 \text{ red}) = \tfrac{8}{125}$$

Note that the total of the four probabilities is 1.

The illustration just discussed is an example of what is known as a **binomial experiment**. A binomial experiment is an experiment for which the following conditions are satisfied:

1. The experiment consists of a fixed number of trials, denoted by *n*.
2. Each trial has only two possible outcomes, usually called 'success' and 'failure'.
3. The outcome of any trial is independent of the outcome of any other trial.
4. The probability of 'success', which we will denote by π, is constant from trial to trial. The probability of failure equals $1 - \pi$, and is also constant from trial to trial. Note that π (the sixteenth letter of the Greek alphabet, pronounced 'pie') does not equal approximately $3 \cdot 14$ in this context, but simply denotes the probability of obtaining a success on any trial.

Each of the *n* trials composing the binomial experiment is called a **binomial trial**.

Let us examine the experiment of drawing the three balls discussed earlier in this section, and note whether it satisfies the four conditions stated above. Considering the conditions in order, we have:

1. Each draw was a trial, and the number was fixed to be 3.
2. Each trial had only two possible outcomes—either a red ball was drawn ('success') or a white ball was drawn ('failure').
3. The outcomes were independent of one another (we assume that after each ball is replaced the balls are remixed).
4. The probability of success is the same on every trial, namely $\tfrac{3}{5}$.

We can consider the drawing of a random sample of size *n* from a population to be a series of *n* trials. If the above four conditions are satisfied, then we will say that we are sampling from a **binomial population**. In particular, when the sample size is fixed, when each member of the sample is either a success or a failure, when whether one member is a success (or failure) does not affect whether any other member is a success (or failure), and when the probability of success is constant from trial to trial, then we say that we are sampling from a binomial population.

Fortunately, there is a general approach to the problem of finding probabilities in situations for which the above conditions hold. We do not need to use a tree diagram, nor do we have to reason our way anew through each different situation. Instead, we will derive a general formula that gives the probability for any specified number of successes, and that is applicable whenever the four conditions given above are true. We will develop this general method after briefly discussing the idea of a random variable.

The number of successes that will result from a binomial experiment is an example of what is known as a **random variable**. That is, it is impossible to predict in advance what number of successes will be the outcome of a binomial experiment because the outcome is due purely to chance. We use **bold-face type** to denote random variables; the number of successes that is the outcome of a binomial experiment is denoted by **x**. In a particular situation we frequently speak of there being x (not bold face) successes, and here x represents a *number*, not a random variable. That is, x is a particular numerical value of the random variable **x**. The probability that the random variable takes on a particular value (say 3, for instance) would be written $P(\mathbf{x} = 3)$. The probability that **x** takes on the value x is written $P(\mathbf{x} = x)$. Again, note that **x** is the symbol for a random variable, while x denotes a particular numerical value of that random variable.

We now return to the task of developing a general formula that gives the probabilities of obtaining the various numbers of successes in n binomial trials. Suppose we have a binomial experiment that is composed of n trials, with probability of success π and probability of failure $1 - \pi$. (The phrase 'binomial experiment' immediately indicates that the trials are independent and that the probability of success is constant from trial to trial.) This situation is completely general, and any formulae that we derive can be applied to any specific problem.

We would like to find the probability that we obtain x successes during the sequence of n binomial trials. In order to find this probability, we will take a roundabout approach. First we will find the probability of getting x successes and $n - x$ failures *in a particular order*. Then, since we are really concerned with the *number* of successes and not with the *order* of the successes, we will find the probability of obtaining x successes *without regard to order*.

One particular order of obtaining x successes and $n - x$ failures is to obtain all of the x successes first, and then the $n - x$ failures,

$$\overbrace{SS\ldots S}^{x}\overbrace{FF\ldots F}^{n-x}$$

Using the conditions that the outcomes are independent and that the probabilities of success and failure remain constant from trial to trial, and using the generalization of Formula (5.9), we have

$$\overbrace{P(SS\ldots S}^{x}\overbrace{FF\ldots F)}^{n-x} = \overbrace{P(S)P(S)\ldots P(S)}^{x}\overbrace{P(F)P(F)\ldots P(F)}^{n-x}$$

$$= \overbrace{\pi\cdot\pi\ldots\pi}^{x}\;\overbrace{(1-\pi)\cdot(1-\pi)\ldots(1-\pi)}^{n-x}$$

$$= \pi^x(1-\pi)^{n-x}$$

Thus, the probability of obtaining x successes and $n-x$ failures, in that order, is $\pi^x(1-\pi)^{n-x}$. When we find the probability of obtaining x successes and $n-x$ failures in any other order, say $SFFS\ldots FSF$, we again have the factor $P(S)$ occurring x times and the factor $P(F)$ occurring $n-x$ times; or, equivalently, we have the factor π occurring x times and the factor $1-\pi$ occurring $n-x$ times. Thus, the probability of obtaining x successes and $n-x$ failures in *any* specified order is $\pi^x(1-\pi)^{n-x}$.

Next we must determine how many specified orders are possible. Think of having n empty numbered spaces which are to be filled by S's and F's.

(1)	(2)	(3)	$\cdots$	(n)

We can make a selection of x of the n positions in which to write S's. After we have written the letter S in each of the x selected spaces we can write the letter F in each of the remaining spaces; this gives us a particular order of S's and F's. Then the number of different orders of successes and failures is equivalent to the number of sets of x objects that can be selected from n objects. (It might be helpful to think of n balls numbered 1 to n in a box. We will select a set of x balls and write the letter S in the spaces that correspond to the numbers on the balls that we draw.) The number of sets of x objects that can be selected from n objects is the number of combinations of n objects taken x at a time, and is denoted C_x^n, or $\binom{n}{x}$, or $C(n,x)$. The symbol $\binom{n}{x}$ will be used in the following discussion.

We are now able to find the probability that we set out to find. We have seen that the probability of x successes and $n-x$ failures in any specific order is $\pi^x(1-\pi)^{n-x}$. The number of different specific orders of x successes and $n-x$ failures, each with the same probability, is $\binom{n}{x}$. Therefore, we have

$$P(x \text{ successes and } n-x \text{ failures}) = \binom{n}{x}\pi^x(1-\pi)^{n-x}$$

To summarize: the probability of obtaining x successes in a binomial experiment consisting of n trials, where π is the probability of success on a single trial, will be denoted by $b(x;\,n,\,\pi)$ and is given by

$$b(x;\,n,\,\pi) = \binom{n}{x}\pi^x(1-\pi)^{n-x} \qquad (5.24)$$

where x can take any value from 0 to n, inclusive.

The quantities n and π are known as *parameters*; they completely specify the binomial distribution that is applicable in any particular situation. Every binomial distribution has probabilities that can be found by substituting the proper values of n and π in Formula (5.24). The binomial distribution is an example of a **discrete distribution,** so called because the random variable x can take on only isolated, discrete values—0, 1, 2, and so on. A theoretical probability distribution can be graphed as a histogram, with the values of the random variable x along the horizontal axis and the values of the probability along the vertical axis.

EXAMPLE: Find the probabilities for all x-values for the binomial distribution with parameters $n = 4$ and $\pi = \frac{1}{3}$. Then sketch the histogram for the distribution.

The formula that gives the probabilities for this particular binomial distribution is

$$b\left(x; 4, \frac{1}{3}\right) = \binom{4}{x}\left(\frac{1}{3}\right)^{x}\left(\frac{2}{3}\right)^{4-x}$$

There are five possible values of x: 0, 1, 2, 3, and 4. To find the required probabilities we successively substitute 0, 1, 2, 3, and 4 into the formula above in place of x. The computations are shown below and the values of x with their corresponding probabilities are shown in Table 5.3.

$$P(x = 0) = b\left(0; 4, \frac{1}{3}\right) = \binom{4}{0}\left(\frac{1}{3}\right)^{0}\left(\frac{2}{3}\right)^{4} = 1 \cdot 1 \cdot \frac{16}{81} = \frac{16}{81} = 0\cdot198$$

$$P(x = 1) = b\left(1; 4, \frac{1}{3}\right) = \binom{4}{1}\left(\frac{1}{3}\right)^{1}\left(\frac{2}{3}\right)^{3} = 4 \cdot \frac{1}{3} \cdot \frac{8}{27} = \frac{32}{81} = 0\cdot395$$

$$P(x = 2) = b\left(2; 4, \frac{1}{3}\right) = \binom{4}{2}\left(\frac{1}{3}\right)^{2}\left(\frac{2}{3}\right)^{2} = 6 \cdot \frac{1}{9} \cdot \frac{4}{9} = \frac{24}{81} = 0\cdot296$$

$$P(x = 3) = b\left(3; 4, \frac{1}{3}\right) = \binom{4}{3}\left(\frac{1}{3}\right)^{3}\left(\frac{2}{3}\right)^{1} = 4 \cdot \frac{1}{27} \cdot \frac{2}{3} = \frac{8}{81} = 0\cdot099$$

$$P(x = 4) = b\left(4; 4, \frac{1}{3}\right) = \binom{4}{4}\left(\frac{1}{3}\right)^{4}\left(\frac{2}{3}\right)^{0} = 1 \cdot \frac{1}{81} \cdot 1 = \frac{1}{81} = 0\cdot012$$

Table 5.3

x	$b(x; n, \pi)$
0	0·198
1	0·395
2	0·296
3	0·099
4	0·012

As a check, we can see whether the five probabilities we found total 1 because, as usual, the total probability associated with a number of mutually exclusive, exhaustive events must be 1.

There are various ways to graph a binomial distribution such as the one displayed in Table 5.3. Perhaps the best way, mathematically speaking, is that shown in Fig. 5.4 (*a*). However, it is easier for the reader to note each probability and to compare it easily with the others if we erect perpendiculars from the x-axis to the dots, forming what we will call a 'discrete line graph', as in Fig. 5.4 (*b*).

A third way to display a binomial distribution is in a histogram similar to those we used to display frequency distributions in Chapter II. Instead of using a line (a length) to represent each probability, we use a rectangle (an area). The rectangle is centred at the x-value, whose probability it represents. The idea of representing a probability by an area is a very important one, and we will return to it in nearly all of the succeeding chapters. The histogram for the binomial distribution of this example is shown in Fig. 5.4 (c).

The binomial distribution is applicable to a wide variety of practical problems, as well as to problems of a more artificial sort, such as those dealing with cards, dice, and coloured balls in boxes. The mathematical Formula (5.24) was derived for an idealized situation. However, the practical situations to which we apply the binomial distribution seldom fit the ideal situation exactly. That is, in the real world to which we hope to apply our statistical knowledge we can seldom be certain that the four conditions necessary for application of the binomial distribution are exactly satisfied. When we have a fixed number of trials and only two outcomes for each trial in a particular situation these facts will be obvious; but frequently we must *assume* that the trials are independent, and that the probabilities are the same from trial to trial.

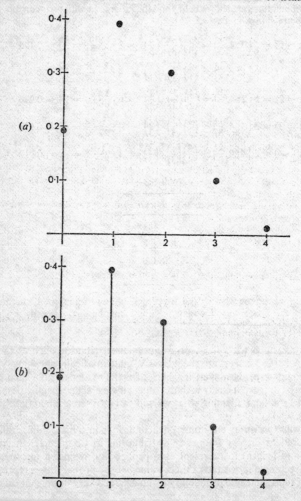

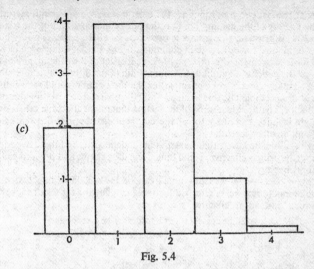

Fig. 5.4

EXAMPLE: Find the probability that in three rolls of a pair of dice exactly one total of 7 is rolled.

Considering each roll as a trial, this experiment is composed of three trials. We note that each trial has two outcomes, rolling a total of 7 (which we will call a 'success') and rolling a total other than 7 (which we will call a 'failure'). If we assume that the trials are all independent and that the probability of rolling a 7 is the same from trial to trial—both of these assumptions are quite reasonable—then this experiment satisfies all the conditions of a binomial experiment, and we can use the binomial distribution to find any desired probabilities.

From the discussion in the section on Probabilities of Two Events we know that the probability of rolling a 7 on any trial is $\frac{1}{6}$. We have

$$P(\text{7 on any roll}) = P(\text{success}) = \pi = \tfrac{1}{6}$$
$$1 - \pi = 1 - \tfrac{1}{6} = \tfrac{5}{6}$$

and
$$n = 3$$

Then the particular binomial distribution which is applicable here is

$$b\left(x; 3, \frac{1}{6}\right) = \binom{3}{x}\left(\frac{1}{6}\right)^x \left(\frac{5}{6}\right)^{3-x}$$

where $x = 0, 1, 2, 3$. The probability of obtaining one success is found by substituting $x = 1$ into the formula immediately above; then we have

$$P(\text{one success}) = b\left(1; 3, \frac{1}{6}\right) = \binom{3}{1}\left(\frac{1}{6}\right)^1 \left(\frac{5}{6}\right)^2$$

$$P(\text{one success}) = \frac{25}{72}$$

EXAMPLE: Over a long period of time a certain drug has been effective in 30 per cent of the cases in which it has been prescribed. If a doctor is now administering this drug to four patients, what is the probability that it will be effective for at least three of the patients?

We can consider the administration of the drug to each patient as a trial. Thus, this experiment has four trials. There are only two outcomes for each trial: the drug is effective or the drug is not effective. So the first two of the four conditions necessary for an experiment to be a binomial experiment are satisfied.

As usual, the last two conditions are the most troublesome. Various factors might affect the validity of the assumption that the third condition, that of the trials being independent, is satisfied. For instance, two or more of the patients might be in the same hospital ward (or otherwise in communication), so that knowledge of the progress of his fellows might affect the physiological effect of the drug on any particular patient; or the patients might have the same strain of the disease, a strain that makes the drug virtually certain to be effective, or the drug administered may actually be a different drug completely, due to an error by the pharmaceutical supplier, pharmacist, or doctor. These are just some of the factors that might affect the actual situation that we are idealizing in order to use the binomial distribution. The reader can no doubt think of other factors.

We will assume, however, that the trials are independent and that the probability of success (the drug is effective) is the same for each patient, and proceed to find the required probability.

The event 'effective for at least three', can be broken down into two mutually exclusive events (outcomes), 'effective for three or effective for four'. If we use the term 'success' instead of 'effective' we can say

$$P(\text{at least 3 successes}) = P(\text{3 successes or 4 successes})$$
$$= P(\text{3 successes}) + P(\text{4 successes})$$
$$= P(x = 3) + P(x = 4).$$

We can now find $P(x = 3)$ and $P(x = 4)$ separately.

In the present example, since the drug is effective in 30 per cent of the cases, we say that the probability that the drug is effective in any single case is $\frac{3}{10}$,

$$P(\text{success}) = \pi = \tfrac{3}{10}$$

Hence, $1-\pi = \frac{7}{10}$. The number of trials is $n = 4$. The equation of the particular binomial distribution is

$$b\left(x; 4, \frac{3}{10}\right) = \binom{4}{x}\left(\frac{3}{10}\right)^x \left(\frac{7}{10}\right)^{4-x}$$

Hence, we have

$$P(x = 3) + P(x = 4) = \binom{4}{3}\left(\frac{3}{10}\right)^3\left(\frac{7}{10}\right)^1 + \binom{4}{4}\left(\frac{3}{10}\right)^4\left(\frac{7}{10}\right)^0$$
$$= 4 \cdot \frac{27}{1000} \cdot \frac{7}{10} + 1 \cdot \frac{81}{10\,000} \cdot 1$$
$$= \frac{837}{10\,000}$$
$$= 0 \cdot 0837$$

The probability that the drug will be effective for at least three of the four patients is 0·0837. Practically interpreted, this statement means that if the drug is administered to 10 000 sets of four patients, in about 837 of the 10 000 sets it will be effective for at least three patients out of four.

EXAMPLE. A rifle-range competitor scores one hit in every four shots, on the average. Assuming that the binomial distribution is applicable, if he has four shots on a particular day, what is: (a) the probability that he will get exactly one hit; (b) the probability that he will get at least one hit?

Considering a hit as a success, we have

$$\pi = P(\text{success}) = \tfrac{1}{4}$$

and

$$1 - \pi = \tfrac{3}{4}$$

The number of trials is 4, and the probabilities for any number of successes (from 0 to 4) are given by the formula

$$b\left(x; 4, \frac{1}{4}\right) = \binom{4}{x}\left(\frac{1}{4}\right)^x\left(\frac{3}{4}\right)^{4-x}$$

(a) We have

$$P(1 \text{ hit}) = P(x = 1) = \binom{4}{1}\left(\frac{1}{4}\right)^1\left(\frac{3}{4}\right)^3$$

$$= 0.422$$

Note in particular that although he gets one hit in 4 tries, on the average, he is not assured of getting 1 hit each time he fires 4 shots. In fact, on only about 4 of 10 such days will he get exactly 1 hit.

(b) We can proceed in a straightforward manner to find that the rifleman will get at least 1 hit. We have

$$P(\text{at least 1 hit}) = P(1 \text{ or } 2 \text{ or } 3 \text{ or } 4 \text{ hits}),$$

which becomes, since the numbers of hits are mutually exclusive events,

$$P(\text{at least 1 hit}) = P(1) + P(2) + P(3) + P(4)*$$

$$= b\left(1; 4, \frac{1}{4}\right) + b\left(2; 4, \frac{1}{4}\right) + b\left(3; 4, \frac{1}{4}\right) + b\left(4; 4, \frac{1}{4}\right)$$

$$= \frac{108}{256} + \frac{54}{256} + \frac{12}{256} + \frac{1}{256} = \frac{175}{256}$$

Thus, we have four probabilities to calculate. They are simple calculations and do not take long, but there is an alternative approach that will frequently save much time.

*$P(1)$ means the same thing as $P(x = 1)$.

The probability of the competitor getting *at least* one hit can be found by the roundabout approach discussed in the earlier section on Combinations of Three or More Events. The event 'at least one hit in four tries' and the event 'no hits in four tries' are complementary events (mutually exclusive, exhaustive events). Therefore,

$$P(\text{at least 1 hit}) + P(\text{no hits}) = 1$$

and $$P(\text{at least 1 hit}) = 1 - P(\text{no hits})$$

The probability of no hits is given by

$$b\left(0; 4, \frac{1}{4}\right) = \binom{4}{0}\left(\frac{1}{4}\right)^0\left(\frac{3}{4}\right)^4$$

$$= 1 \cdot 1 \cdot \frac{81}{256}$$

$$= 0.316$$

Therefore, $$P(\text{at least 1 hit}) = 1 - 0.316 = 0.684$$

This approach can always be used when we need to find the probability that at least one of a number of events will occur. We simply find the probability that none of the events will occur and use the relationship

$$P(\text{at least 1}) = 1 - P(\text{none})$$

Now try questions 11–18 of the Exercises.

THE THEORETICAL MEAN OF THE BINOMIAL DISTRIBUTION

The theoretical mean and variance of the binomial distribution are needed in Chapters VI and VII. The discussion presented in this chapter is intuitive, not rigorous; the author has attempted to explain the formulae for the mean and variance rather than to prove them.

Consider the binomial distribution with parameters $n = 5$ and $\pi = \frac{1}{2}$. Its formula is

$$b\left(x; 5, \frac{1}{2}\right) = \binom{5}{x}\left(\frac{1}{2}\right)^x\left(\frac{1}{2}\right)^{5-x} = \binom{5}{x}\left(\frac{1}{2}\right)^5$$

The values of $b(x; 5, \frac{1}{2})$ for $x = 0, 1, 2, 3, 4,$ and 5 are found to be $\frac{1}{32}$, $\frac{5}{32}$, $\frac{10}{32}$, $\frac{10}{32}$, $\frac{5}{32}$, $\frac{1}{32}$, respectively. The histogram corresponding to this binomial distribution is shown in Fig. 5.5.

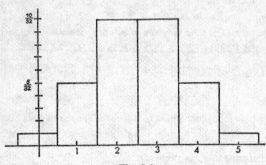

Fig. 5.5

This histogram is a *theoretical* histogram; the probabilities corresponding to the *x*-values are *theoretical* probabilities. If the assumptions underlying the binomial distribution are met, then the probability of four successes (for instance) in five trials is $\frac{5}{32}$. The histograms studied in Chapter II were empirical histograms. They displayed the number of observations that were *observed* to be in each class.

Perhaps the analogy between a histogram representing theoretical *probabilities* and one representing empirical frequencies is somewhat strained. Let us change the probabilities to frequencies so that we will have a histogram representing theoretical frequencies corresponding to another histogram representing empirical frequencies. We can make this change by multiplying each of the six probabilities by 32. This eliminates the fractions and we have the theoretical frequency distribution shown in Table 5.4. The corresponding histogram is shown in Fig. 5.6.

Table 5.4

x-value	Theoretical frequency
0	1
1	5
2	10
3	10
4	5
5	1

Suppose that we perform an experiment which consists of tossing a coin five times and counting the number of heads thrown in the five tosses. If we perform this experiment 32 times, then *theoretically* we would *expect* to get no heads on 1 performance of the experiment, 1 head on 5 performances, 2 heads on 10 performances, and so on. (The observed results don't always match the expected results closely; more will be said about this point later.)

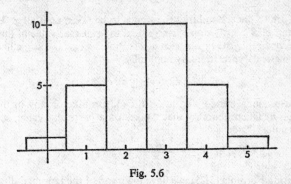

Fig. 5.6

The theoretical frequency distribution in Table 5.4 is exactly analogous to the sample frequency distributions discussed in Chapter II. The mean of a theoretical frequency distribution is called the **theoretical mean**, and is found just as the sample mean is found. We can treat the above theoretical frequency distribution exactly as we would a sample distribution: The class interval is one unit, and the class marks are 0, 1, 2, 3, 4, and 5.

The class marks and the corresponding frequencies of the theoretical frequency distribution are shown in Table 5.4. Recalling that the sample mean for classified data is given by $\bar{x} = \frac{1}{n} \sum x_i' f_i$, we have the theoretical mean, usually denoted by the Greek letter μ (mu) as

$$\mu = \tfrac{1}{32}[(0 \times 1) + (1 \times 5) + (2 \times 10) + (3 \times 10) + (4 \times 5) + (5 \times 1)]$$
$$= \tfrac{1}{32}[0 + 5 + 20 + 30 + 20 + 5]$$
$$= \tfrac{1}{32} \cdot 80 = \tfrac{5}{2}$$

Algebraically, the first line of the above calculation of μ is equivalent to

$$\mu = (0 \times \tfrac{1}{32}) + (1 \times \tfrac{5}{32}) + (2 \times \tfrac{10}{32}) + (3 \times \tfrac{10}{32}) + (4 \times \tfrac{5}{32}) + (5 \times \tfrac{1}{32})$$

When we display the values of x and their corresponding probabilities in a table (see Table 5.5) we see that the last expression for μ is simply the sum of the products found by multiplying each x by its corresponding probability.

Table 5.5

x	0	1	2	3	4	5
$P(x) =$	$\tfrac{1}{32}$	$\tfrac{5}{32}$	$\tfrac{10}{32}$	$\tfrac{10}{32}$	$\tfrac{5}{32}$	$\tfrac{1}{32}$

After having made this observation, it is just a step to generalize: to find the mean of a theoretical discrete (the random variable x can take on only isolated values) distribution, multiply each value of x by its corresponding theoretical probability and add the resulting products. Symbolically

$$\mu = \sum x \cdot P(x) \tag{5.25}$$

where we are to understand that the sum is to be taken over all x. The theoretical mean of a distribution is also called the **expected value of the random variable x**, or simply the **expectation of x**, denoted $E(x)$, and sometimes called the **mathematical expectation**. Symbolically,

$$\mu = E(x) = \sum x \cdot P(x) \tag{5.26}$$

If x is a binomial random variable, then x can take on any of the values 0, 1, 2, ... n. The probability that x takes on a particular value, x, is given by the formula

$$b(x; n, \pi) = \frac{n!}{x!(n-x)!} \pi^x (1 - \pi)^{n-x}$$

Then, using Formula (5.25), we have the mean of the binomial distribution as

$$\mu = \sum_{x=0}^{n} x \cdot b(x; n, \pi)$$

Substituting the quantity to which $b(x; n, \pi)$ is equivalent into the expression for μ immediately above, we have

$$\mu = \sum_{x=0}^{n} x \cdot \frac{n!}{x!(n-x)!} \pi^x (1 - \pi)^{n-x}$$

Using mathematics which is beyond the scope of this book, the last sum can be shown to equal $n\pi$.

To summarize,

$$\mu = n\pi \tag{5.27}$$

is the theoretical mean or population mean of the binomial distribution with parameters n *and* π (n trials with probability of success for each trial equal to π).

The reader should realize that Formula (5.25) is quite general. Although the content of that formula has already been stated once, it is worth repeating: to find the theoretical mean of any discrete distribution, multiply each value of x by its corresponding probability and add the resulting products. Sometimes a theoretical distribution (or probability distribution, as it is frequently called) is specified by means of a formula, as the binomial distribution is. Often, however, there is no formula that will give the probabilities corresponding to the x-values; in such cases the distribution is specified by simply listing the x-values and their probabilities in a table of some sort. An example of such a distribution is illustrated by the discussion that follows.

Recall the dice examples discussed earlier in this chapter. There are eleven possible totals when two dice are thrown—2 through 12; each has a certain probability. Let the random variable x equal the total number of spots showing on a throw of two dice. Then the probability distribution of x is given in Table 5.6.

Table 5.6

$x =$	2	3	4	5	6	7	8	9	10	11	12
$P(x) =$	$\frac{1}{36}$	$\frac{2}{36}$	$\frac{3}{36}$	$\frac{4}{36}$	$\frac{5}{36}$	$\frac{6}{36}$	$\frac{5}{36}$	$\frac{4}{35}$	$\frac{3}{36}$	$\frac{2}{36}$	$\frac{1}{36}$

From Formula (5.26) we have

$$\mu = E(\mathbf{x}) = 2(\tfrac{1}{36}) + 3(\tfrac{2}{36}) + 4(\tfrac{3}{36}) + \ldots + 11(\tfrac{2}{36}) + 12(\tfrac{1}{36})$$
$$= \frac{2 + 6 + 12 + 20 + 30 + 42 + 40 + 36 + 30 + 22 + 12}{36}$$
$$= \tfrac{252}{36} = 2\tfrac{1}{3} = 7$$

Saying that 7 is the expected value of **x** does not mean that we 'expect' to throw a 7 every time—it means that we 'expect' to throw a total of 7 *on the average*, and, generally speaking, the more times we toss a pair of dice, the closer to 7 the average number of spots will be. Frequently, in fact, the expected value of **x** will equal a number that **x** itself cannot possibly equal. In a prior example the expected number of heads in 5 tosses of a coin was found to be $\tfrac{5}{2}$, a number that the number of heads on a particular sequence of 5 tosses cannot possibly equal. But in the long run we expect the *average* number of heads per 5 tosses to become very close to $\tfrac{5}{2}$.

THE THEORETICAL VARIANCE OF THE BINOMIAL DISTRIBUTION

In Formula (4.16) we defined the sample variance for classified data to be given by

$$s^2 = \frac{1}{n-1} \sum_{i=1}^{k} (x'_i - \bar{x})^2 f_i$$

Moving the $\dfrac{1}{n-1}$ behind the summation, we obtain

$$s^2 = \sum_{i=1}^{k} (x'_i - \bar{x})^2 \frac{f_i}{n-1} \tag{5.28}$$

Once again, in order to obtain a formula for the theoretical variance, we draw analogies between empirical quantities and theoretical quantities.

The values of the binomial random variable **x** are analogous to the empirical class marks x'_i.

The theoretical probabilities $P(x)$ are analogous to the empirical 'relative frequencies' $\dfrac{f_i}{n-1}$. (The f_i would need to be divided by n, rather than $n-1$, to actually be a relative frequency.)

The theoretical mean μ is analogous to the empirical mean $\bar{x}$.

When we replace the empirical quantities in Formula (5.28) by their theoretical counterparts, we obtain a formula for the theoretical variance, usually denoted σ^2, which is analogous to the one for the sample variance:

$$\sigma^2 = \sum (x - \mu)^2 P(x) \tag{5.29}$$

where the sum is to be taken over all values of **x**.

Another formula for the theoretical variance, which can be derived from Formula (5.29), is

$$\sigma^2 = \sum x^2 P(x) - \mu^2 \qquad (5.30)$$

with the summation again being taken over all possible x.

EXAMPLE: Consider the distribution of x, the number of heads in five tosses of a coin, that we have discussed previously:

$x =$	0	1	2	3	4	5
$P(x) =$	$\frac{1}{32}$	$\frac{5}{32}$	$\frac{10}{32}$	$\frac{10}{32}$	$\frac{5}{32}$	$\frac{1}{32}$

Find the theoretical variance of x (the variance of the distribution). Peform the calculations two ways, by applying Formula (5.29) and by applying Formula (5.30).

You recall from the discussion in the previous section that the mean for this theoretical distribution is $\mu = \frac{5}{2}$ (Formula (5.27)).

The computations necessary to find σ^2 from Formula (5.29) are shown in Table 5.7.

Table 5.7

x	$x - \mu$	$(x - \mu)^2$	$P(x)$	$(x - \mu)^2 P(x)$
0	$-\frac{5}{2}$	$\frac{25}{4}$	$\frac{1}{32}$	$\frac{25}{128}$
1	$-\frac{3}{2}$	$\frac{9}{4}$	$\frac{5}{32}$	$\frac{45}{128}$
2	$-\frac{1}{2}$	$\frac{1}{4}$	$\frac{10}{32}$	$\frac{10}{128}$
3	$\frac{1}{2}$	$\frac{1}{4}$	$\frac{10}{32}$	$\frac{10}{128}$
4	$\frac{3}{2}$	$\frac{9}{4}$	$\frac{5}{32}$	$\frac{45}{128}$
5	$\frac{5}{2}$	$\frac{25}{4}$	$\frac{1}{32}$	$\frac{25}{128}$

We add the entries in the last column to get the necessary sum, and obtain

$$\sigma^2 = \sum_{x=0}^{5} (x - \mu)^2 P(x) = \frac{160}{128} = \frac{5}{4}$$

The computations are somewhat shorter if Formula (5.30) is used to find σ^2. The necessary computations are shown in Table 5.8.

Table 5.8

x	x^2	$P(x)$	$x^2 P(x)$
0	0	$\frac{1}{32}$	0
1	1	$\frac{5}{32}$	$\frac{5}{32}$
2	4	$\frac{10}{32}$	$\frac{40}{32}$
3	9	$\frac{10}{32}$	$\frac{90}{32}$
4	16	$\frac{5}{32}$	$\frac{80}{32}$
5	25	$\frac{1}{32}$	$\frac{25}{32}$

Adding the entries in the last column, we have

$$\sum_{x=0}^{5} x^2 P(x) = \frac{240}{32}$$

Recalling that $\mu = \frac{5}{2}$ and applying Formula (5.30), we have

$$\sigma^2 = \sum x^2 P(x) - \mu^2$$
$$= \frac{240}{32} - \left(\frac{5}{2}\right)^2 = \frac{240}{32} - \frac{25}{4} = \frac{240}{32} - \frac{200}{32} = \frac{5}{4}$$

Formulas (5.29) and (5.30) are quite general and can be applied to any theoretical discrete distribution for which the x-values and their corresponding probabilities are given (or if we are given the possible x-values and a formula that yields their probabilities).

More particularly, when we have a binomial distribution with parameters n and π, Formula (5.30) becomes

$$\sigma^2 = \left[\sum_{x=0}^{n} x^2 \cdot \binom{n}{x} \pi^x (1-\pi)^{n-x} \right] - (n\pi)^2 \qquad (5.31)$$

The sum $\sum_{x=0}^{n} x^2 \binom{n}{x} \pi^x (1-\pi)^{n-x}$ can be shown to equal $n\pi + n(n-1)\pi^2$ by methods beyond the scope of this book. Then (5.31) yields

$$\begin{aligned}
\sigma^2 &= [n\pi + n(n-1)\pi^2] - (n\pi)^2 \\
&= n\pi + n^2\pi^2 - n\pi^2 - n^2\pi^2 \\
&= n\pi - n\pi^2 \\
&= n\pi(1-\pi)
\end{aligned}$$

In summary: the theoretical variance of a binomial distribution with parameters n and π is given by the formula

$$\sigma^2 = n\pi(1-\pi) \qquad (5.32)$$

In the above problem $n = 5$, $\pi = \frac{1}{2}$. $\therefore$ $\sigma^2 = 5 \cdot \frac{1}{2} \cdot \frac{1}{2} = \frac{5}{4}$ as already found.

EXAMPLE: Find the theoretical variance of the binomial distribution whose parameters are $n = 50$ and $\pi = \frac{2}{5}$.

The variance is

$$\begin{aligned}
\sigma^2 &= n\pi(1-\pi) \\
&= 50(\tfrac{2}{5})(\tfrac{3}{5}) \\
&= 12
\end{aligned}$$

EXAMPLE: The probability that a certain rifleman will get a hit on any given shot at the rifle range is $\frac{3}{10}$. If he fires one hundred shots, find the theoretical mean and variance of x, the number of hits. Assume that the binomial distribution is applicable.

We have $n = 100$, $\pi = \frac{3}{10}$, and $1 - \pi = \frac{7}{10}$. Then, by Formula (5.27), the mean is

$$\mu = n\pi = 100(\tfrac{3}{10}) = 30$$

and by equation (5.32) the variance is

$$\sigma^2 = n\pi(1-\pi) = 100(\tfrac{3}{10})(\tfrac{7}{10}) = 21$$

Note how easy it is to find the mean of a binomial distribution now that we have a simple formula at our disposal. It takes much less time to find a variance from Formula (5.32) than from (5.29) or (5.30). Keep in mind, however, that (5.29) and (5.30) are applicable to *any distribution*, whereas (5.32) is applicable *only to the binomial distribution*.

EXERCISES

1. A single card is drawn from an ordinary pack of playing cards. What is the probability that the card is:

(a) an ace?
(b) a five?
(c) a face card?
(d) a red card?
(e) a club?

2. There are fifteen slips of paper in a hat, numbered from 1 to 15. If one slip is drawn at random, find the probability that

(a) the number drawn is 5.
(b) the number drawn is even.
(c) the number drawn is odd.
(d) the number drawn is divisible by 3.

3. Two dice are rolled. Find the probability of rolling:

(a) a total of 5;
(b) a total greater than 8;
(c) a total less than or equal to 7.

4. Two events, A and B, are mutually exclusive: $P(A) = \frac{1}{5}$, and $P(B) = \frac{1}{3}$. Find the probability that:

(a) either A or B will occur;
(b) both A and B will occur;
(c) neither A nor B will occur.

5. Two events, A and B, are independent: $P(A) = 0\cdot3$, and $P(B) = 0\cdot5$.

(a) Find $P(AB)$.
(b) Find $P(A + B)$.
(c) Find the probability that neither A nor B will occur.

6. Two cards are drawn in succession (without replacement) from an ordinary pack of playing cards. Find the probability that:

(a) the cards are both red;
(b) the cards are both black;
(c) the cards are the same colour;
(d) the cards are different colours.

7. Find the probability of drawing three successive non-face cards in three successive draws (without replacement) from a pack of 52 cards.

8. Find the probability that each of the cards of question 7 has a number of spots divisible by 4.

9. Find the probability of throwing one of the following totals on a single throw of a pair of dice: a total of 2, 7, 11.

10. What is the probability of throwing a 7 followed by 11 in two consecutive throws of a pair of dice?

11. Evaluate: (a) $5!$; (b) $6!/3!$; (c) $_6P_3$; (d) $_7P_2$; (e) $_6C_1$; (f) $_8C_3$.

12. If no letter can be used more than once: (a) How many three-letter arrangements can be made from the letters A, B, C, D, E? (b) How many four-letter arrangements? (c) How many five-letter arrangements?

13. In a certain classroom there are six desks in the front row. If there are fifteen students in the class, how many different arrangements are possible for the front row? (Assume that no empty seats are to be left on the front row.)

14. Thornton has 25 books. He can take only four of them on a visit to his grandparents' house. How many different sets of four books might he select to take from the 25?

15. How many different five-card poker hands are there? (Do not compute.)

16. A club has a membership of six men and four women. A three-person committee is chosen at random. What is the probability that it consists: (a) entirely of men; (b) of two men and one woman?

17. Find: (a) $b(4; 6, \frac{1}{2})$; (b) $b(2; 6, \frac{1}{3})$; (c) $b(4; 7, \frac{1}{4})$.

18. A card is drawn from an ordinary pack of playing cards, and its suit (clubs,

diamonds, hearts, spades) noted; then it is replaced, the pack is shuffled, and another card is drawn. This is done until four cards have been drawn.

(a) What is the probability that two spades will be drawn in four draws?

(b) What is the probability that at least two spades will be drawn in four draws?

(c) What is the probability that two red cards will be drawn in four draws?

(d) What is the probability that at most two red cards will be drawn in four draws?

19. Determine the probability of getting: (a) exactly three heads in 6 tosses of a fair coin; and (b) at least three heads in 6 tosses of a fair coin.

20. Determine the variance for the distribution of x, the number of heads in 6 tosses of a coin by using Formulae (5.30) and (5.32).

CHAPTER VI

THE NORMAL DISTRIBUTION

INTRODUCTION

The preceding chapter introduced the most important discrete distribution, the binomial distribution. The present chapter deals with the most important continuous distribution, the normal distribution. Among other topics, the relationship between the binomial and normal distributions is also discussed.

A discrete random variable is a variable which can take on only isolated values. These values are nearly always positive integers because we nearly always deal with experiments in which we *count*. For instance, let us consider the situation outlined in the last example in Chapter V. Assuming that a rifleman firing one hundred shots is a binomial experiment with probability of success on any trial equal to $\frac{3}{10}$, then the random variable x, the number of successes in one hundred trials, has a binomial distribution with parameters $n = 100$ and $\pi = \frac{3}{10}$. On any particular performance of the experiment, the random variable x has some integral value between 0 and 100, inclusive, with values around 30 being most common. That is, the number of hits in 100 shots must be a whole number between 0 and 100; x is a discrete variable.

Some other examples of discrete variables are:

(1) the number of boys in families containing six children;
(2) the number of strawberries picked from a plant during a season;
(3) the number of colds people catch during the winter;
(4) the number of puppies per litter for dogs;
(5) the number of peas per dozen pods which contain maggots.

The reader can doubtless furnish many more examples. All of these examples of discrete variables have something in common—they are *counts* of something —boys, strawberries, colds, puppies, and peas. Hence, whenever we have a counting situation, we know that the random variable involved will be discrete.

There are many situations, however, in which we do not count—we *measure*. We can find measurements as accurately as we please if our measuring instruments are sufficiently precise. For example, usually heights of individuals are measured and recorded to the nearest unit. Given a precise measuring instrument and method of measuring, we could, if we wished, find heights correct to the nearest tenth of a unit, hundredth of a unit, or to any specified degree of accuracy. A height which is given as 67 units might be found to be 67·3 units, and if measured more accurately, 67·29 units, and yet more accurately, 67·288 units, and so on. Theoretically, then, a person's height might be *any* value and not just isolated values such as 67, 68, or 69 units.

The preceding discussion motivates us to define a **continuous random variable** as a variable which might theoretically take on *any* value in the range of possible values. Some illustrations of continuous variables are:

(1) the weights of individuals;
(2) the distances which various cars of a certain make will travel on a litre of petrol;
(3) the lengths of the right feet of individuals;
(4) the times that it takes light bulbs of a certain brand to burn out.

To summarize the distinction between discrete and continuous variables: **a discrete variable can take on only isolated values, usually positive integral values, whereas a continuous variable can theoretically take on any value.** More practically, a discrete variable results when something is *counted*; a continuous variable results when something is *measured*—time, weight, height, liquid volume, distance, and so on.

THE NORMAL DISTRIBUTION

There are a great number of continuous distributions. The normal distribution is undoubtedly the one that is the most widely used in applications of statistics, and is the one with which we will deal at the greatest length in this book. Several other continuous distributions that we will discuss later are the t-distribution, the χ^2 distribution, and the F-distribution. [χ, written 'chi' and pronounced 'ki' (rhyming with sky), is the twenty-second letter of the Greek alphabet.]

A normal distribution is completely specified by two parameters, the theoretical mean and the theoretical variance of the population. If a random variable **x** has a normal distribution, with mean μ and variance σ^2, then the equation of the distribution is

$$f(x) = \frac{1}{\sigma\sqrt{2\pi}} \cdot e^{-\frac{1}{2}\left(\frac{x-\mu}{\sigma}\right)^2}$$

Do not be intimidated by this equation. The number π should be familiar to you from elementary geometry, or perhaps from trigonometry or elsewhere; it is approximately equal to 3·1416. The number e is an important constant in mathematics, although you may not be acquainted with it. The number e is the base of natural logarithms and occurs frequently in various other roles in mathematics (calculus and beyond); it is approximately equal to 2·7183.

Although this equation will be referred to several times in succeeding sections, it will not be used to work problems, and the reader does not need to use it. There is a different normal curve for each combination of μ and σ^2, but they all have the same characteristics. The graph of any normal distribution is a symmetric, rather bell-shaped curve which slopes downward on both sides from the maximum value (which always occurs at the point where $x = \mu$, incidentally) towards the x-axis, but never touches the x-axis. The normal distribution for which $\mu = 5$ and $\sigma^2 = 4$ has

$$f(x) = \frac{1}{2\sqrt{2\pi}} \cdot e^{-\frac{1}{2}\left(\frac{x-5}{2}\right)^2}$$

as its equation; the graph of this normal distribution is shown in Fig. 6.1. One reason the normal distribution is so important is that a number of natural phenomena (that is, the *measurements* of these phenomena) are normally distributed or nearly so. Phenomena such as heights and weights of

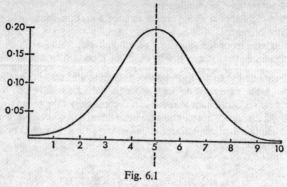

Fig. 6.1

individuals, I.Q. scores, errors in measuring the length of a metal rod with high accuracy, and scores on mathematics tests all have distributions that are normal. Practically speaking, this means that if we select a sample of 100 people and measure their weights, then classify these observations and draw the histogram, the histogram will follow roughly the outlines of a normal curve. Perhaps the histogram (and corresponding normal curve) would look like the one in Fig. 6.2.

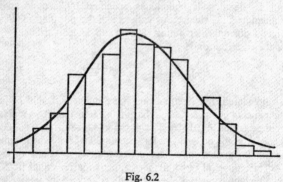

Fig. 6.2

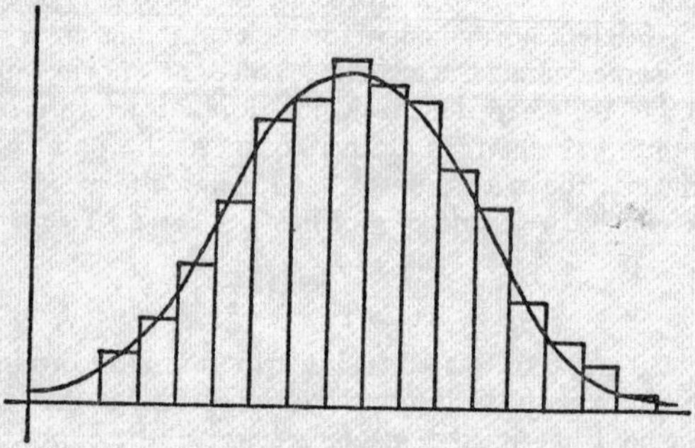

Fig. 6.3

When we take a larger sample, then the histogram follows the shape of a normal distribution much more faithfully. If our sample size were five hundred, then perhaps the histogram would look like the one in Fig. 6.3, which has very close agreement between histogram and the normal curve.

USE OF STANDARD NORMAL TABLES

A random variable x is said to have been **standardized** when it has been adjusted so that its mean is 0 and its standard deviation is 1. Standardization can be effected by subtracting μ, the mean of x, from x, and dividing the resulting difference by σ, the standard deviation of x: thus $\frac{x - \mu}{\sigma}$ is a standardized variable.

One of the most important theorems of mathematical statistics, which we will not attempt to prove in this book, states that if x is a normal random variable with mean μ and standard deviation σ, then the standardized variable $\frac{x - \mu}{\sigma}$ has the normal distribution with mean 0 and standard deviation 1. The normal distribution with mean 0 and standard deviation 1 is known as the **standard normal distribution**. The variable $\frac{x - \mu}{\sigma}$ is known as the **standard normal variable**, and is usually denoted by z.

Areas under the standard normal curve are frequently needed and are therefore widely tabulated, filling only a couple of pages. Having a method by which *any* normal curve can be changed into a single normal curve is a tremendous space-saver and time-saver. If there were no such method there would need to be an infinite number of tables to take care of all possible combinations of μ and σ. (μ can be any real number, and σ can be any positive real number.) If tables were not available we would need to use techniques from calculus to find necessary areas. Even if the reader has studied calculus, he probably cannot evaluate the definite integral

$$\int_{4\cdot523}^{4\cdot867} \frac{1}{3\sqrt{2\pi}}\, e^{-\frac{1}{2}\left(\frac{x - 4\cdot523}{3}\right)^2}\, dx$$

It should be evident that finding areas by this method would be a difficult and time-consuming procedure. The standard normal tables are a tremendous

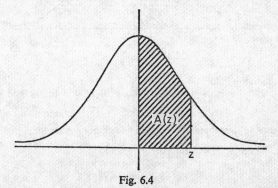

Fig. 6.4

labour-saver, and you should be appreciative of this fact each time you use them (see page 230).

Let us denote by $A(z)$ the area underneath the standard normal from 0 to z, where z is any number, positive, negative, or zero. Most standard normal tables give the values of $A(z)$, shown in Fig. 6.4, for positive values of z in steps of $0\cdot01$ from $z = 0$ to $z = 4\cdot00$ or $z = 5\cdot00$.

EXAMPLE: Find the area under the standard normal curve from 0 to $1\cdot93$.

Turning to the standard normal table in the Appendix, we scan down the z-column until we find $1\cdot93$. Opposite $1\cdot93$ we read the value of $A(1\cdot93)$, $0\cdot47320$, in the $A(z)$ column. Thus, we have

$$A(1\cdot93) = 0\cdot47320$$

In order to gain some facility in using the standard normal table, the reader can verify the $A(z)$ values given below:

$$A(0\cdot03) = 0\cdot01197$$
$$A(0\cdot67) = 0\cdot24857$$
$$A(2\cdot57) = 0\cdot49492$$
$$A(3\cdot21) = 0\cdot49934$$

The standard normal curve is symmetric about the line perpendicular to the z-axis through the point where $z = 0$. Crudely interpreted, this means that if the curve were cut out of the page and folded along the line $z = 0$, then the two halves of the curve would match perfectly. Since the total area under the curve must equal 1 (as in Chapter V, the total probability—which here is an area—must equal 1), the area under the curve to the left of $z = 0$ equals $\frac{1}{2}$ and the area under the curve to the right of $z = 0$ equals $\frac{1}{2}$. The symmetry of the curve also allows us to find $A(z)$ for negative values of z, as shown in Fig. 6.5.

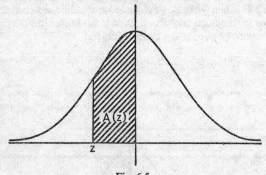

Fig. 6.5

EXAMPLE: Find $A(-1\cdot50)$.

We know that $A(-1\cdot50) = A(1\cdot50)$. We find $A(1\cdot50)$ as before, and have

$$A(-1\cdot50) = 0\cdot43319$$

Some other values of $A(z)$ for negative values of z are:

$$A(-0\cdot75) = 0\cdot27337$$
$$A(-1\cdot64) = 0\cdot44950$$
$$A(-3\cdot00) = 0\cdot49865$$

The area under the standard normal curve to the left of z will be denoted by the symbol $\Phi(z)$. For positive z, as shown in Fig. 6.6, the area underneath the

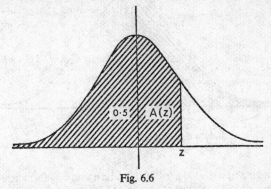

Fig. 6.6

curve to the left of z equals the area under the curve to the left of 0 plus the area under the curve from 0 to z. We have seen that the area under the curve to the left of 0 is $\frac{1}{2}$, and we can find the area under the curve from 0 to z, $A(z)$, from the table. Thus, if z is positive,

$$\Phi(z) = 0.50000 + A(z) \tag{6.1}$$

The reader can verify the following values of $\Phi(z)$ for $z = 1.82$, $z = 1.96$, and $z = 2.58$. Sketches are frequently helpful, and this should be kept in mind in all of these examples of the use of the standard normal tables. We find

$$\Phi(1.82) = 0.50000 + 0.46562 = 0.96562$$
$$\Phi(1.96) = 0.50000 + 0.47500 = 0.97500$$
$$\Phi(2.58) = 0.50000 + 0.49506 = 0.99506$$

If z is negative, then $\Phi(z)$ is an area similar to that shaded in Fig. 6.7.

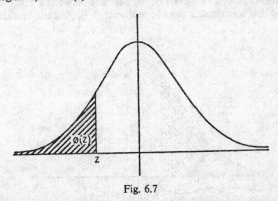

Fig. 6.7

The value of $\Phi(z)$ may be found as follows: We know that the area to the left of 0 is $\frac{1}{2}$. The area from 0 to z is $A(z)$ and can be found. By inspection

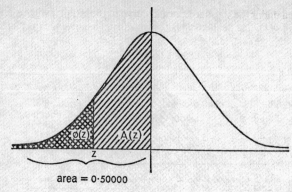

area = 0·50000

Fig. 6.8

of Fig. 6.8 we see that the value of $\Phi(z)$ can be found by subtracting $A(z)$ from 0·50000. Thus, if z is negative, the value of $\Phi(z)$ is given by the relation

$$\Phi(z) = 0·50000 - A(z) \qquad (6.2)$$

EXAMPLE: Find $\Phi(-0·45)$.

We first need to find $A(-0·45)$. We know that $A(-0·45) = A(0·45)$ which has been found to be 0·17364. Thus $A(-0·45) = 0·17364$. Substituting this value into Formula (6.2), we have

$$\Phi(-0·45) = 0·50000 - 0·17364 = 0·32636$$

Several more values of $\Phi(z)$ for negative z are

$$\Phi(-1·82) = 0·50000 - A(-1·82) = 0·50000 - 0·46562 = 0·03438$$
$$\Phi(-1·96) = 0·50000 - A(-1·96) = 0·50000 - 0·47500 = 0·02500$$
$$\Phi(-2·58) = 0·50000 - A(-2·58) = 0·50000 - 0·49506 = 0·00494$$

A frequently encountered problem in probability and statistics is that of finding the probability that the random variable **z** lies between two values a and b, written $P(a < z < b)$. When we say '**z** is between a and b' it is to be understood that a is smaller than b, written $a < b$. You may remember that,

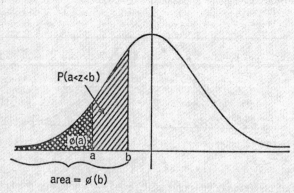

area = $\phi(b)$

Fig. 6.9

unless $a < b$, it is incorrect to write $a < z < b$ (which, in words, means a is less than z and z is less than b). Geometrically interpreted, $P(a < z < b)$ equals the area under the curve from a to b in Fig. 6.9. In order to find the area under the curve from a to b, we can proceed as follows:

(1) find the area under the curve to the left of b, $\Phi(b)$;
(2) find the area under the curve to the left of a, $\Phi(a)$;
(3) from Fig. 6.9 we see that $\Phi(b) - \Phi(a)$ gives us the required area.

Hence, we have shown that $P(a < z < b)$ is given by

$$P(a < z < b) = \Phi(b) - \Phi(a) \tag{6.3}$$

EXAMPLE: Find $P(-0.47 < z < 0.94)$.
Using the standard normal table we find

$$A(0.94) = 0.32639$$
and
$$A(-0.47) = A(0.47) = 0.18082$$

From Formula (6.1) we have

$$\Phi(0.94) = 0.50000 + 0.32639 = 0.82639$$

and from Formula (6.2)

$$\Phi(-0.47) = 0.50000 - 0.18082 = 0.31918$$

Substituting these results in Formula (6.3), we obtain

$$P(a < z < b) = \Phi(0.94) - \Phi(-0.47)$$
$$= 0.82639 - 0.31918$$
$$= 0.50721$$

The method illustrated above, which makes use of Formula (6.3), is quite general. Formula (6.3) and the procedure illustrated in the above example can be used for any values of a and b. There are alternative methods that can be used to find $P(a < z < b)$, with different methods being needed when a and b are both positive, when a and b are both negative, and when a is negative and b is positive. We will show the alternative procedure for the preceding example, which is the case where a is negative and b is positive, in the following example. The reader can draw sketches of the other two possible cases, and develop and use procedures alternative to the more general procedure if he wishes.

EXAMPLE: Find $P(-0.47 < z < 0.94)$. See Fig. 6.10.

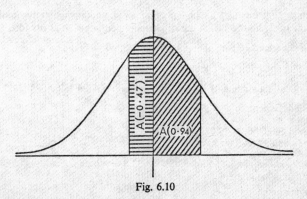

Fig. 6.10

The probability that z lies between -0.47 and 0.94 is equal to the area shaded in Fig. 6.10. We note that the shaded area is the sum of two separate areas, $A(-0.47)$ and $A(0.94)$. Therefore

$$P(a < z < b) = A(-0.47) + A(0.94)$$
$$= 0.18082 + 0.32639$$
$$= 0.50721$$

which, naturally, agrees with the result we obtained previously.

The two probabilities, which the reader may verify, are

$$P(0.17 < z < 0.29) = \Phi(0.29) - \Phi(0.17)$$
$$= 0.61409 - 0.56750$$
$$= 0.04659$$
and
$$P(-0.89 < z < -0.03) = \Phi(-0.03) - \Phi(-0.89)$$
$$= 0.48803 - 0.18673$$
$$= 0.30130$$

We frequently wish to find the probability that z exceeds a certain number, say a, which is written $P(z > a)$. Geometrically interpreted, $P(z > a)$ is the area under the curve to the right of a in Fig. 6.11.

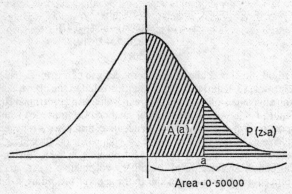

Fig. 6.11

We know that the area to the *left* of a is $\Phi(a)$, which we can find easily. We also know that the total area under the curve equals 1. It is obvious that

(Area left of a) + (Area right of a) = (Total area)

Solving this equation for (Area right of a) we have

(Area right of a) = (Total area) − (Area left of a)

which is equivalent to $\qquad P(z > a) = 1 - \Phi(a) \qquad$ (6.4)

EXAMPLE: Find $P(z > 1.00)$.
Using Formula (6.4), we have

$$P(z > 1.00) = 1 - \Phi(1.00)$$
$$= 1 - 0.84134$$
$$= 0.15866$$

NOTE: If the number a is positive, $P(z > a)$ is an area such as that shaded in Fig. 6.11, and can be found by subtracting $A(a)$ from 0·50000.

If the number a is negative $P(z > a)$ is found by adding $A(a)$ to 0·50000 (the reader can supply a sketch).

Both of the above two possible situations are taken care of by the single formula $P(z > a) = 1 - \Phi(a)$. The reader may use whichever method he prefers, the general or the particular.

MORE NORMAL PROBABILITIES

The knowledge of how to use the standard normal tables is valuable indeed, but seldom are we fortunate enough to work with a random variable that has a standard normal distribution. Nearly always we will deal with random variables, x, that are merely normal, not standard normal, and we will need to find probabilities such as $P(x < a)$, $P(x > a)$, and $P(a < x < b)$. We cannot find such probabilities immediately, but we can use what we have already learned to work them into forms that we are equipped to handle. The necessary technique is that of standardizing the normal random variable x; then we know how to handle a standard normal variable.

To briefly repeat what has previously been said about standardizing and about standardized normal variables: we standardize any random variable by subtracting its mean from itself and dividing the resulting difference by its standard deviation. If x is a normal variable, with mean μ and variance σ^2, then the variable

$$z = \frac{x - \mu}{\sigma}$$

is a standard normal variable.

Incidentally, a convenient shorthand notation for the statement 'x is a normal random variable with mean μ and variance σ^2' is $x = N(\mu, \sigma^2)$. Thus $x = N(55, 25)$ means that x has a normal distribution with mean 55 and variance 25. The statement made in the last sentence of the preceding paragraph can be written compactly: If $x = N(\mu, \sigma^2)$, then $z = \frac{x - \mu}{\sigma} = N(0, 1)$.

In order to find $P(x < a)$, where $x = N(\mu, \sigma^2)$, we standardize by setting

$$z = \frac{x - \mu}{\sigma} \tag{6.5}$$

so that we can employ the standard normal distribution. Solving Formula (6.5) for x, we obtain

$$x = \mu + \sigma z \tag{6.6}$$

When x is replaced by $\mu + \sigma z$, the inequality $x < a$ becomes

$$\mu + \sigma z < a \tag{6.7}$$

To solve (6.7) for z we first subtract μ from both sides (the same quantity can be subtracted from, or added to, both sides of an inequality), obtaining

$$\sigma z < a - \mu \tag{6.8}$$

and then multiply both sides of (6.8) by $\frac{1}{\sigma} \left(\frac{1}{\sigma} \right.$ is positive because σ is positive by definition, and if both sides of an inequality are multiplied by the same positive number the direction of the inequality sign is unchanged) and obtain

$$z < \frac{a - \mu}{\sigma} \qquad (6.9)$$

Thus, whenever x has a value less than some number a, z has a value less than $\frac{a - \mu}{\sigma}$. This statement is equivalent to saying that

$$P(x < a) = P\left(z < \frac{a - \mu}{\sigma}\right)$$

In the previous section we denoted the probability that z is less than any number, say k, by $\Phi(k)$. Using this notation, we have

$$P(x < a) = P\left(z < \frac{a - \mu}{\sigma}\right) = \Phi\left(\frac{a - \mu}{\sigma}\right) \qquad (6.10)$$

EXAMPLE: Given that x has a normal distribution with mean 10 and standard deviation 4, find $P(x < 15)$. From Formula (6.10) we have

$$P(x < 15) = P\left(z < \frac{15 - 10}{4}\right)$$
$$= P(z < 1\cdot25)$$
$$= \Phi(1\cdot25)$$

From tables,
$$\Phi(1\cdot25) = 0\cdot50000 + A(1\cdot25)$$
$$= 0\cdot50000 + 0\cdot39435$$
$$= 0\cdot89435$$

The required probability is $0\cdot8944$.

In addition to having to find $P(x < a)$, we also frequently need to find $P(x > a)$ and $P(a < x < b)$. The procedure that was used to obtain Formula (6.10) is used once again to obtain formulae for each of the probabilities $P(x > a)$ and $P(a < x < b)$.

Whenever $x > a$, we know that $z > \frac{a - \mu}{\sigma}$. Hence we have

$$P(x > a) = P\left(z > \frac{a - \mu}{\sigma}\right)$$

which becomes (making use of Formula (6.4))

$$P(x > a) = P\left(z > \frac{a - \mu}{\sigma}\right) = 1 - \Phi\left(\frac{a - \mu}{\sigma}\right) \qquad (6.11)$$

EXAMPLE: If $x = N(9,9)$, find $P(x > 2)$.
Noting that the standard deviation is $\sqrt{9} = 3$, and using Formula (6.11), we have

$$P(x > 2) = 1 - \Phi\left(\frac{2 - 9}{3}\right)$$
$$= 1 - \Phi(-2\cdot33)$$
$$= 1 - 0\cdot00990$$
$$= 0\cdot99010$$

EXAMPLE: The true masses of ten kilogramme sacks of potatoes processed at a certain packing house have a normal distribution with mean 10 (kilogrammes) and variance 0·01 (square kilogrammes). What is the probability that a sack purchased at the greengrocer's shop will have a mass of at least 9·875 kg?

Let x = number of kilogrammes of potatoes in a 10-kg sack. We have $\mu = 10$ and $\sigma = \sqrt{0·01} = 0·1$, and we want to find $P(x > 9·875)$.

From Formula (6.11) we have

$$P(x > 9·875) = P\left(z > \frac{9·875 - 10}{0·1}\right) = 1 - \Phi\left(\frac{9·875 - 10}{0·1}\right)$$
$$= 1 - \Phi(-1·25) = 1 - 0·10565$$
$$= 0·89435$$

Whenever $a < x < b$, we have $\dfrac{a - \mu}{\sigma} < z < \dfrac{b - \mu}{\sigma}$;

therefore $\qquad P(a < x < b) = P\left(\dfrac{a - \mu}{\sigma} < z < \dfrac{b - \mu}{\sigma}\right)$

Making use of Formula (6.3), we can say

$$P(a < x < b) = P\left(\frac{a - \mu}{\sigma} < z < \frac{b - \mu}{\sigma}\right)$$
$$= \Phi\left(\frac{b - \mu}{\sigma}\right) - \Phi\left(\frac{a - \mu}{\sigma}\right) \quad (6.12)$$

EXAMPLE: If x has a normal distribution with mean 9 and standard deviation 3, find $P(5 < x < 11)$.

Using Formula (6.12) above, we have

$$P(5 < x < 11) = P\left(\frac{5 - 9}{3} < z < \frac{11 - 9}{3}\right)$$
$$= P(-1·33 < z < 0·66)$$
$$= \Phi(0·66) - \Phi(-1·33)$$
$$= 0·74537 - 0·09176$$
$$= 0·65361$$

THE NORMAL APPROXIMATION TO THE BINOMIAL

Whenever the number of trials in a binomial experiment is small it is easy to find the probabilities of the various values of x, the number of successes, by using the formula

$$b(x; n, \pi) = \binom{n}{x}\pi^x(1 - \pi)^{n-x}$$

As the number of trials increases, however, the effort involved in answering questions about probabilities associated with the experiment quickly becomes laborious.

For instance, suppose that we want to know the probability that in fifteen tosses of a coin we toss at least nine heads. You will undoubtedly agree that

$n = 15$ is not a particularly large number of trials. However, in order to find $P(x \geqslant 9)$ we say

$$P(x \geqslant 9) = P(x = 9 \text{ or } 10 \text{ or } 11 \text{ or } 12 \text{ or } 13 \text{ or } 14 \text{ or } 15)$$
$$= b\left(9; 15, \frac{1}{2}\right) + \ldots + b\left(15; 15, \frac{1}{2}\right)$$

So we have seven probabilities to compute, after which we must perform the addition. This isn't particularly difficult, but it takes a fair amount of time. To find only one of these probabilities, for example, we have

$$P(x = 10) = b\left(10; 15, \frac{1}{2}\right) = \binom{15}{10}\left(\frac{1}{2}\right)^{10}\left(\frac{1}{2}\right)^{5}$$

$$\binom{15}{10} = \frac{15!}{10!5!} = \frac{15 \cdot 14 \cdot 13 \cdot 12 \cdot 11}{1 \cdot 2 \cdot 3 \cdot 4 \cdot 5} = 3003$$

therefore
$$\left(\frac{1}{2}\right)^{10}\left(\frac{1}{2}\right)^{5} = \left(\frac{1}{2}\right)^{15} = \frac{1}{32\,768};$$

$$\binom{15}{10}\left(\frac{1}{2}\right)^{10}\left(\frac{1}{2}\right)^{5} = 3003\left(\frac{1}{32\,768}\right) = 0 \cdot 092$$

correct to three places.

Thus, you see that if we were to calculate the other six such probabilities we would expend a considerable amount of time and energy. And this is a very simple problem.

Suppose that we had been asked to find the probability of throwing at least 55 heads in 90 tosses of a coin. Then we would need to calculate 36 probabilities, involving such factorial expressions as $\frac{90!}{55!35!}$. Although there are tables which give binomial probabilities for various combinations of values n and π, there are no such tables in this book. Although none are included in this book, tables of binomial coefficients are also readily available, especially for values of n no larger than 20. Or, if in the binomial coefficient $\binom{n}{r}$, n is so large that $\binom{n}{r}$ is not in the table of binomial coefficients, tables of $n!$ and $\log n!$ can be used to evaluate $\binom{n}{r}$. Tables of factorials and their logarithms can be found in some textbooks, in collections of mathematical tables, and in other sources.

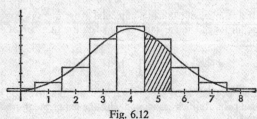

Fig. 6.12

Fortunately, there is a method that enables us to find the approximate value of binomial probabilities with a relatively small amount of computation. This method is known as the **normal approximation to the binomial**. A normal curve is fitted to the histogram of a binomial distribution. If π is close to $\frac{1}{2}$, then the theoretical histogram of the binomial distribution is rather symmetrical and n need not be very large in order for the histogram to be such that it can be approximated very nicely by a normal curve. For instance, let $\pi = \frac{1}{2}$ and $n = 8$. The histogram of the binomial distribution is shown in Fig. 6.12, along with the normal curve that approximates it.

If π is not close to $\frac{1}{2}$ and n is small, then the fit of the normal curve to the binomial histogram is not very good. For example, the histogram in Fig. 6.13 is for $\pi = \frac{1}{10}$ and $n = 8$. It is obvious that a normal curve will not fit this histogram satisfactorily. If π is not close to $\frac{1}{2}$, then n must be fairly large in order to fit a normal curve to the histogram.

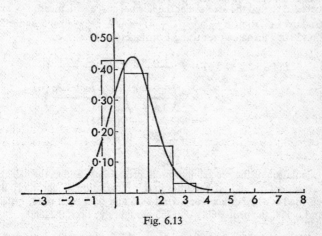

Fig. 6.13

When we examine Fig. 6.12 it seems reasonable to say that the normal curve should have the same mean as that of the binomial distribution that it approximates. It also seems reasonable to want the standard deviation of the normal curve to equal that of the binomial distribution.

Recall that the mean of the binomial distribution is $n\pi$, and that its standard deviation is $\sqrt{n\pi(1 - \pi)}$. Then we will use the normal distribution with mean $n\pi$ and standard deviation $\sqrt{n\pi(1 - \pi)}$ to approximate the binomial distribution. The approximation improves as the binomial distribution 'spreads out', for it can be shown that as the number of trials n increases, the shape of the histogram of the binomial distribution becomes more and more like that of the normal curve.

This assertion is important and useful enough to be stated more formally as a theorem.

Theorem. If x represents the number of successes in n independent trials of an experiment for which π is the probability of success in a single trial, then the variable $\dfrac{x - n\pi}{\sqrt{n\pi(1 - \pi)}}$ has a distribution that approaches the standard normal distribution as the number of trials becomes increasingly large.

This theorem is proved in mathematical statistics courses by advanced mathematical methods. Essentially, it *guarantees* that if n is sufficiently large we can use the normal distribution with mean $n\pi$ and standard deviation $\sqrt{n\pi(1-\pi)}$ to approximate quite accurately the binomial distribution.

To illustrate the procedure for using the normal distribution to approximate probabilities associated with the binomial distribution, we will consider the binomial distribution with parameters $n = 8$ and $\pi = \frac{1}{2}$ (see Fig. 6.12). We will find the approximate probability that $x = 5$ (the number of successes is 5). The exact probability, of course, is the area of the small rectangle centred at $x = 5$.

The approximate probability is the area bounded by the normal curve, the x-axis, and the lines which form the upright sides of the rectangle centred at the point on the x-axis whose co-ordinate is 5. This area is shaded in Fig. 6.12. Thus, we need to find the area under the normal curve with mean $\mu = n\pi = 4$ and standard deviation $\sigma = \sqrt{n\pi(1-\pi)} = \sqrt{2}$, between 4·5 and 5·5. We have found such areas before: using Formula (6.12) we have

$$P(4\cdot5 < x < 5\cdot5) = P\left(\frac{4\cdot5 - 4}{\sqrt{2}} < z < \frac{5\cdot5 - 4}{\sqrt{2}}\right)$$
$$= \Phi\left(\frac{5\cdot5 - 4}{\sqrt{2}}\right) - \Phi\left(\frac{4\cdot5 - 4}{\sqrt{2}}\right)$$
$$= \Phi(1\cdot061) - \Phi(0\cdot354)$$
$$= 0\cdot85566 - 0\cdot63833$$
$$= 0\cdot21733$$

In the calculation of the preceding normal probability, and in the calculation of all subsequent normal probabilities in this illustration, linear interpolation has been used for higher accuracy. Had 1·06 and 0·35 been used, instead of 1·061 and 0·354, the probability 0·21733 would have been 0·21860.

NOTE: When an irrational number such as $\sqrt{2}$ is in the denominator the easiest way to perform the computation is to 'rationalize the denominator'. The detailed computation below will illustrate:

$$\frac{5\cdot5 - 4}{\sqrt{2}} = \frac{1\cdot5}{\sqrt{2}}$$
$$= \frac{1\cdot5}{\sqrt{2}} \cdot \frac{\sqrt{2}}{\sqrt{2}} \quad \text{(multiplying numerator and denominator by } \sqrt{2}\text{)}$$
$$= \frac{1\cdot5 \cdot \sqrt{2}}{\sqrt{4}} \quad (\sqrt{2} \cdot \sqrt{2} = \sqrt{4})$$
$$= \frac{1\cdot5(1\cdot414)}{2} \quad \text{(looking up } \sqrt{2} \text{ in tables)}$$
$$= 1\cdot061 \quad \text{(performing computation)}$$

Usually all the intermediate steps would be skipped. One simply writes

$$\frac{5\cdot5 - 4}{\sqrt{2}} = \frac{1\cdot5(1\cdot414)}{2} = 1\cdot061$$

An alternative procedure is simply to replace $\sqrt{2}$ by its approximate value, 1·414, and divide:

$$\frac{1\cdot5}{\sqrt{2}} = \frac{1\cdot5}{1\cdot414} = 1\cdot061$$

The probability that the binomial random variable **x** equals any other specified value (except the 'end' ones, 0 and n, which will be treated separately in a moment) is found in the same manner. The general procedure (and formulae) follows.

Let $P_B(\mathbf{x} = a)$ represent the probability that the binomial random variable **x** equals a:

$$P_B(\mathbf{x} = a) = b(a; n, \pi)$$

Let
$$P_N(a - \tfrac{1}{2} < \mathbf{x} < a + \tfrac{1}{2})$$

represent the probability that the normal random variable **x** lies between $a - \tfrac{1}{2}$ and $a + \tfrac{1}{2}$.

Then, whenever the binomial random variable **x** is not equal to 0 or n, we have

$$P_B(\mathbf{x} = a) \cong P_N(a - \tfrac{1}{2} < \mathbf{x} < a + \tfrac{1}{2})$$

NOTE: The symbol $\cong$ means 'approximately equal to'.

Using Formula (6.12), we obtain

$$P_B(\mathbf{x} = a) \cong P_N\left(\frac{(a - \tfrac{1}{2}) - n\pi}{\sqrt{n\pi(1 - \pi)}} < \mathbf{z} < \frac{(a + \tfrac{1}{2}) - n\pi}{\sqrt{n\pi(1 - \pi)}}\right) \quad (6.13)$$

Thus,
$$P_B(\mathbf{x} = 2) \cong P_N\left(\frac{3}{2} < \mathbf{x} < \frac{5}{2}\right)$$
$$= P\left(\frac{\tfrac{3}{2} - 4}{\sqrt{2}} < \mathbf{z} < \frac{\tfrac{5}{2} - 4}{\sqrt{2}}\right)$$
$$= \Phi\left(\frac{\tfrac{5}{2} - 4}{\sqrt{2}}\right) - \Phi\left(\frac{\tfrac{3}{2} - 4}{\sqrt{2}}\right)$$
$$= \Phi(-1 \cdot 061) - \Phi(-1 \cdot 768)$$
$$= 0 \cdot 14434 - 0 \cdot 03853$$
$$= 0 \cdot 10581$$

All the area under the normal curve to the *left* of $\tfrac{1}{2}$ is taken as the approximate probability that the random variable **x** equals 0. (Some authors use the area under the normal curve between $-\tfrac{1}{2}$ and $\tfrac{1}{2}$ as the approximate probability that $\mathbf{x} = 0$. The reason why we prefer the alternative is explained later.) Hence we have

$$P_B(\mathbf{x} = 0) \cong P_N(\mathbf{x} < \tfrac{1}{2}) \quad (6.14)$$
$$= P_N\left(\mathbf{z} < \frac{\tfrac{1}{2} - n\pi}{\sqrt{n\pi(1 - \pi)}}\right)$$

In the present illustration,

$$P(\mathbf{x} = 0) = P\left(\mathbf{z} < \frac{\tfrac{1}{2} - 4}{\sqrt{2}}\right)$$
$$= \Phi\left(\frac{\tfrac{1}{2} - 4}{\sqrt{2}}\right)$$
$$= \Phi(-2 \cdot 475)$$
$$= 0 \cdot 00666$$

All the area under the normal curve to the *right* of $n - \tfrac{1}{2}$ is taken as the approximate probability that the binomial random variable **x** equals n. (Some authors use the area under the normal curve between $n - \tfrac{1}{2}$ and $n + \tfrac{1}{2}$. We explain why we prefer the alternative after the following illustration.) Hence we have

$$P_B(\mathbf{x} = n) \cong P_N(\mathbf{x} > n - \tfrac{1}{2}) = P_N\left(z > \frac{(n - \tfrac{1}{2}) - n\pi}{\sqrt{n\pi(1 - \pi)}}\right) \quad (6.15)$$

In the present illustration,

$$\begin{aligned}
P_B(\mathbf{x} = 8) &\cong P_N(\mathbf{x} > 7\tfrac{1}{2}) \\
&= P\left(z > \frac{7\tfrac{1}{2} - 4}{\sqrt{2}}\right) \\
&= 1 - \Phi\left(\frac{7\tfrac{1}{2} - 4}{\sqrt{2}}\right) \\
&= 1 - \Phi(2\cdot475) \\
&= 1 - 0\cdot99334 \\
&= 0\cdot00666
\end{aligned}$$

We use all the area under the curve to the left of $\tfrac{1}{2}$ to approximate $P_B(\mathbf{x} = 0)$ and all the area under the curve to the right of $n - \tfrac{1}{2}$ to approximate $P_B(\mathbf{x} = n)$, so that all the total approximate probabilities will sum to 1. The probabilities would not sum to 1 if we took the area under the curve from $\tfrac{1}{2}$ to $-\tfrac{1}{2}$ to give the approximate probability that $\mathbf{x} = 0$ and the area under the curve from $n - \tfrac{1}{2}$ to $n + \tfrac{1}{2}$ to give the approximate probability that $\mathbf{x} = n$.

We now have the approximate probabilities for the x-values 0, 2, 5, and 8. The computations for these values, along with those for the x-values 1, 3, 4, 6, and 7, are summarized below:

$$\begin{aligned}
P(\mathbf{x} = 0) &\cong \Phi(-2\cdot475) = 0\cdot00666 \\
P(\mathbf{x} = 1) &\cong \Phi(-1\cdot768) - \Phi(-2\cdot475) = 0\cdot03187 \\
P(\mathbf{x} = 2) &\cong \Phi(-1\cdot061) - \Phi(-1\cdot768) = 0\cdot10581 \\
P(\mathbf{x} = 3) &\cong \Phi(-0\cdot354) - \Phi(-1\cdot061) = 0\cdot21733 \\
P(\mathbf{x} = 4) &\cong \Phi(0\cdot354) - \Phi(-0\cdot354) = 0\cdot27666 \\
P(\mathbf{x} = 5) &\cong \Phi(1\cdot061) - \Phi(0\cdot354) = 0\cdot21733 \\
P(\mathbf{x} = 6) &\cong \Phi(1\cdot768) - \Phi(1\cdot061) = 0\cdot10581 \\
P(\mathbf{x} = 7) &\cong \Phi(2\cdot475) - \Phi(1\cdot768) = 0\cdot03187 \\
P(\mathbf{x} = 8) &\cong 1 - \Phi(2\cdot475) = 0\cdot00666
\end{aligned}$$

Note that the total of these nine probabilities is $1\cdot00000$. This furnishes a convenient check.

In order to compare the approximate probabilities with the true probabilities, we must calculate the true binomial probabilities.

In a binomial experiment with eight trials, and probability of success $\tfrac{1}{2}$, the probability of x successes in eight trials is

$$b\left(x; 8, \frac{1}{2}\right) = \binom{8}{x}\left(\frac{1}{2}\right)^x \left(\frac{1}{2}\right)^{8-x} = \binom{8}{x}\left(\frac{1}{2}\right)^8$$

Correct to 6 places we can find

$$\left(\frac{1}{2}\right)^8 = 0\cdot003906$$

That we have $\qquad b\left(x; 8, \frac{1}{2}\right) = \binom{8}{x}(0\cdot003906)$

and the binomial probabilities are:

$$P(x = 0) = \binom{8}{0}(0.003906) = 1 . (0.003906) = 0.00391$$

$$P(x = 1) = \binom{8}{1}(0.003906) = 8 . (0.003906) = 0.03125$$

$$P(x = 2) = \binom{8}{2}(0.003906) = 28(0.003906) = 0.10937$$

$$P(x = 3) = \binom{8}{3}(0.003906) = 56(0.003906) = 0.21874$$

$$P(x = 4) = \binom{8}{4}(0.003906) = 70(0.003906) = 0.27342$$

As can be seen when the definition of $\binom{n}{r}$ is applied (see Formula (5.23)), $\binom{8}{5} = \binom{8}{3}$, $\binom{8}{6} = \binom{8}{2}$, $\binom{8}{7} = \binom{8}{1}$, and $\binom{8}{8} = \binom{8}{0}$. Therefore, the last four probabilities are the same as the first four (in reverse order):

$$P(x = 5) = 0.21874$$
$$P(x = 6) = 0.10937$$
$$P(x = 7) = 0.03125$$
and
$$P(x = 8) = 0.00391$$

NOTE: In general, $\binom{n}{r} = \binom{n}{n-r}$.

Although $n = 8$ is rather small, the normal curve furnishes a very satisfactory approximation if $\pi = \frac{1}{2}$, which gives a symmetric binomial distribution. Excepting the extreme values of 0 and 8, no approximate value is in error more than about 3 per cent. The relative error of the approximation for the values 0 and n is about 70 per cent, although the absolute error is rather small, only about three-thousandths.

When n is as small as 8 a value of π as small as $\frac{1}{10}$, say, would result in very poor agreement between the binomial distribution and the approximating normal curve (see Fig. 6.13). The binomial distribution would be quite unsymmetrical with most of the area close to zero, whereas the normal distribution would be symmetrical and centred at $\frac{4}{5}$.

In order to avoid attempting to fit a normal curve to a histogram which is very unsymmetrical, with most of the area piled up at one end or the other, some conditions need to be met before the normal approximation can be employed. The usual conditions are: if π is less than or equal to $\frac{1}{2}$, then $n\pi$ must be at least 5; if π is greater than or equal to $\frac{1}{2}$, then $n(1 - \pi)$ must be at least 5. Symbolically, the conditions are

$$n\pi \geqslant 5 \quad \text{if} \quad \pi \leqslant \tfrac{1}{2}$$
and
$$n(1 - \pi) \geqslant 5 \quad \text{if} \quad \pi \leqslant \tfrac{1}{2}$$

If these two conditions are satisfied, then we are guaranteed that the mean of the binomial distribution (and that of the approximating normal distribution, as well) will be at least 5 units from the closest end (either 0 or n). Thus, a reasonably good fit of the normal distribution to the binomial distribution is assured.

EXAMPLE: A pair of dice is thrown 120 times. What is the approximate probability: of throwing at least 15 sevens; of throwing between 20 and 30 sevens, inclusive?

The rolls are independent. The probability of rolling a seven is constant from trial to trial. Also, each roll results in either a success or a failure. Therefore the number of sevens tossed is a binomial random variable.

To find the probability of at least 15 sevens we want to find $P_B(x \geqslant 15)$.

$$n = 120, \ \pi = \tfrac{1}{6}, \ n\pi = 20, \ \sqrt{n\pi(1 - \pi)} = 4 \cdot 08248$$

$$P_B(x \geqslant 15) \cong P_N \left(z > \frac{14 \cdot 5 - 20}{4 \cdot 0825} \right) \quad \text{(Formula (6.15))}$$

$$= P(z > -1 \cdot 35)$$

$$= 1 - \Phi(-1 \cdot 35) = 1 - 0 \cdot 0885 = 0 \cdot 9115$$

The probability of throwing between 20 and 30 sevens, inclusive, is given by

$$P_B(20 \leqslant x \leqslant 30) \cong P_N \left(\frac{19 \cdot 5 - 20}{4 \cdot 0825} < z < \frac{30 \cdot 5 - 20}{4 \cdot 0825} \right)$$

$$= P(-0 \cdot 12 < z < 2 \cdot 57)$$

$$= \Phi(2 \cdot 57) - \Phi(-0 \cdot 12)$$

$$= 0 \cdot 9949 - 0 \cdot 4522 = 0 \cdot 5427$$

EXAMPLE: Suppose that in a large city it is desired to get an estimate of the proportion of voters in favour of a certain proposal by taking a sample of 200. What is the probability that a majority of the persons in the sample are against the proposal if, in reality, only 45 per cent of the electorate is against the proposal.

Note that we must assume that the population is sufficiently large so that the probability of 'success' on each trial is constant, and that the trials are independent. We have $n = 200$ and $\pi = 0 \cdot 45$.

We want to find $P_B(x \geqslant 101)$, which is given by

$$P_B(x > 100) = P_N \left(z > \frac{100 \cdot 5 - 90}{7 \cdot 03562} \right) = 1 - \Phi(1 \cdot 49)$$

$$= 1 - 0 \cdot 9319$$

$$= 0 \cdot 068$$

EXERCISES

1. Find $A(z)$ for each of the following values of z:

    (a) 1·75   (b) 0·09   (c) 2·33   (d) 1·23   (e) 0·77

2. Find $A(z)$ for each of the following values of z:

    (a) −0·68   (b) −0·99   (c) −1·25   (d) −2·58   (e) −1·45

3. Find $\Phi(z)$ for each of the z-values in Exercise 1.

4. Find $\Phi(z)$ for each of the z-values in Exercise 2.

5. Given that z is a standard normal variable. Find:

    (a) $P(0 \cdot 25 < z < 1 \cdot 63)$ (b) $P(1 \cdot 02 < z < 1 \cdot 37)$

    (c) $P(-1 \cdot 75 < z < 0 \cdot 14)$ (d) $P(-2 \cdot 00 < z < -1 \cdot 00)$

    (e) $P(-0 \cdot 43 < z < 0 \cdot 82)$ (f) $P(-1 \cdot 33 < z < 1 \cdot 33)$

    (g) $P(z > 1 \cdot 15)$ (h) $P(z > -0 \cdot 50)$

    (i) $P(z < -1 \cdot 23)$

6. Given that x is a normal variable with mean 50 and variance 100, find:

    (a) $P(x < 60)$ (b) $P(x < 58 \cdot 7)$

    (c) $P(x > 48 \cdot 2)$ (d) $P(x > 45)$

    (e) $P(45 \cdot 0 < x < 65 \cdot 3)$ (f) $P(42 \cdot 3 < x < 58 \cdot 0)$

7. Given that **x** is a normal variable with mean 4·42 and variance 1·96, find:

  (*a*) $P(4·00 < x < 5·00)$ (*b*) $P(x > 6·00)$

  (*c*) $P(x < 5·83)$ (*d*) $P(3·50 < x < 4·82)$

8. If a pair of dice are thrown, what is the approximate probability of throwing: (*a*) at least 15 sevens in 120 rolls; (*b*) not more than 30 sevens?

9. Given that the random variable **x** has a binomial distribution with $n = 100$ and $\pi = \frac{1}{2}$. Find the approximate value of each of the following probabilities, using the normal approximation to the binomial.

  (*a*) $P(x > 40)$ (*b*) $P(x \geqslant 47)$

  (*c*) $P(x < 45)$ (*d*) $P(51 < x < 55)$

  (*e*) $P(48 < x < 52)$

10. A rifleman gets 3 hits out of every 10 shots on the average. Assuming that the probability that he gets a hit is constant from shot to shot, and that each shot is independent of every other shot, find the approximate probability that he will get at least 40 hits in his next 100 shots.

11. What is the approximate probability of throwing less than 35 heads in 100 tosses of an unbiased coin?

12. What is the approximate probability of throwing between 47 and 53 heads in 100 tosses of an unbiased coin?

CHAPTER VII

SOME TESTS OF STATISTICAL HYPOTHESES

INTRODUCTION

Chapters VII and VIII are devoted entirely to the testing of statistical hypotheses. In addition, portions of Chapters IX, XI, XII, and XIII deal with hypothesis-testing, and Chapter X, in which confidence intervals are discussed, depends heavily on material in Chapters VII, VIII, and IX.

In the hypothesis-testing situations in Chapters VII, VIII, and IX, we will test a hypothesis about a theoretical quantity whose value is unknown. This theoretical quantity (or population quantity) is known as a **parameter** (we have already used this term in Chapter V). The hypothesis about the population quantity will be tested by means of a sample quantity (or empirical quantity) known as a **statistic**. A statistic is some quantity which is calculated from the observations composing a sample. This process of making a decision about a theoretical population quantity on the basis of an observed sample quantity is known as **statistical inference**. We use an observed quantity to infer something about an unknown theoretical quantity; a sample quantity to infer something about a population quantity; a statistic to infer something about a parameter.

THE NATURE OF A STATISTICAL HYPOTHESIS— TWO TYPES OF ERROR

In order to present the ideas involved in testing a statistical hypothesis, and to discuss the two types of errors which can be made, a typical hypothesis-testing situation will be discussed at some length in this section.

Consider the following experimental situation: a certain disease (which is uncomfortable, but not serious) can be treated with a certain drug, and experience has shown that within three days 50 per cent of those to whom the drug was administered show no symptoms of the disease whatsoever. A second drug has been developed, and preliminary experimentation has shown that it might be more effective than the first drug. In order to make a decision about the effectiveness of the second drug, a medical research worker plans to have the drug administered to 100 patients. The number of patients who show no symptoms of the disease at the end of three days will be counted, and on the basis of this number a decision will be made about the effectiveness of the second drug relative to the first drug.

Clearly, common sense indicates that the second drug would not be judged more effective than the first unless more than 50 of the 100 patients recovered within three days. But the question is: how many more than 50 are necessary? Are 51 recoveries necessary? 55? 60? 75? Just how many patients must recover before it is certain that the second drug is better? Unfortunately, the methods of statistics do not yield certain answers. In any given situation it is impossible to make a decision which is *known* to be correct; instead, a decision which is *likely* to be correct is made. And whenever a decision is made, there is a possibility of error.

Suppose, for example, that 65 of the 100 patients recovered within three days. Then the research worker is faced with two possible decisions:

1. The second drug is, in reality, no better than the first, even though 65 out of 100 recovered. It is unlikely that as many as 65 would recover if the second drug is no better than the first, although it could happen due merely to chance. I choose to believe that the second drug is, in reality, no better than the first, even though 65 out of 100 recovered.

2. I realize that as many as 65 out of 100 could recover merely by chance, even though the second drug is no better than the first. However, such a possibility is very unlikely. Having observed 65 recoveries, it seems more reasonable for me to believe that the second drug *is* more effective than the first.

Keep in mind that the true state of affairs is unknown to the research worker. He has observed a particular result of an experiment and wants to select the most rational alternative. If he selects the first of the two alternatives above—that the second drug is no more effective than the first—and if the second drug actually *is* more effective, then he has made an error. (We will see later that this sort of error is known as a Type II error.) On the other hand, if the worker selects the second alternative—that the second drug is more effective than the first—and if the second drug actually *is not* more effective, then he has again made an error. (We will see later that this sort of error is known as a Type I error.)

We do not yet have a criterion by which a rational decision can be made. First of all, some assumptions need to be made. Then we can use our knowledge of the binomial and normal distributions to formulate a criterion for making a decision.

We will assume that the administration of the drug to 100 patients constitutes 100 trials of a binomial experiment. That is, the trials are independent and the probability of success is constant from trial to trial. (Obviously the number of trials is fixed and there are only two outcomes, recovery within three days or not, for each trial.) If the second drug is as effective as the first the number of successes (recoveries within three days) has a binomial distribution with parameters $\pi = 0.5$ and $n = 100$. If the expected number of successes is significantly greater than 50 the research worker is going to conclude that the second drug is better. But how should the word 'significantly' be interpreted? This word must be spelt out in quantitative terms.

The research worker can select a certain (small) number (let us say 0.05 for the sake of definiteness in the following discussion), and use the following rule: if the number of recoveries that is observed has a probability of occurring (calculated on the assumption that the second drug is no better than the first) which is less than the selected number (0.05, say), then we will conclude that the second drug is better than the first.

Note that even if the second drug is no better than the first, 5 per cent of the time a number of recoveries would be observed, which by the rule just stated would lead the research worker to conclude that the second drug *is* better.

From Chapter V it is known that the mean and variance of a binomial distribution with $n = 100$ and $\pi = 0.5$ are 50 and 25 respectively. From

Chapter VI it is known that a normal curve fits a binomial distribution quite well. We want to find the x-value on the fitted normal curve that is exceeded by 5 per cent of the x-values. If a number of recoveries greater than this x-value is observed, then the worker will conclude that the second drug is better. We know that 5 per cent of the area under the curve lies to the right of the point which is 1·645 standard deviations above the mean. The mean is $n\pi = 50$. The standard deviation is $\sqrt{n\pi(1 - \pi)} = 5$. Therefore, 5 per cent of the area lies to the right of the point that is (1·645)(5) units greater than 50, i.e. the point 58·23. Therefore the worker will conclude that the second drug is superior to the first if he observes 59 or more successes.

More generally, the hypothesis that is being tested is called the **null hypothesis,** and is denoted by H_o. The hypothesis that the experimenter is willing to accept if he does not accept the null hypothesis is called the **alternative hypothesis,** and is usually denoted by H_1.

Our null hypothesis—that the second drug is no better than the first—is equivalent to stating that the true proportion of successes in the binomial experiment is no greater than 0·50. Symbolically,

$$H_o: \pi \leqq 0\cdot50$$

The alternative hypothesis—that the second drug is better than the first—is equivalent to stating that the proportion of successes is greater than 0·50. Symbolically,

$$H_1: \pi > 0\cdot50$$

These two hypotheses can be equivalently stated in terms of the expected *number* of successes, rather than in terms of the theoretical (true) value of π. The hypotheses would then be written

$$H_o: n\pi \leqq 50$$
$$H_1: n\pi > 50$$

The arbitrarily selected probability (in the preceding discussion it was 0·05) is denoted by α (alpha, the first letter of the Greek alphabet) and is called the **α-level of the test.** The α determines an interval, and if the experiment yields a value of the quantity which we are using to test the null hypothesis within this interval we reject the null hypothesis. For this reason the interval is called the **rejection region** or the **critical region.** (In the preceding discussion the rejection region was any value of x greater than 58·23.) The value of α is the probability of rejecting H_o when H_o is, in fact, true. That is, α is the size of the Type I error, which is also known as the α-error. Using the idea of conditional probability, we can say that the Type I error is the probability of rejecting H_o, given that H_o is true. Symbolically,

$$\text{Size of Type I error} = P(\text{rejecting } H_o \mid H_o \text{ true})$$

A Type II error is made when H_o is erroneously accepted—when H_o is accepted even though it is false. (H_o is not *known* to be false, of course, or it would not be accepted; bear in mind that the truth or falsity of H_o is unknown, even after it is accepted or rejected.) The size of the Type II error (the probability of making a Type II error) is denoted by β (beta, the second letter of the Greek alphabet) and the Type II error is often called the β-error. The

Type II error is the probability of accepting H_o given that H_o is false. Symbolically,

$$\text{Size of Type II error} = P(\text{accepting } H_o \mid H_o \text{ false})$$

Just to summarize briefly:

Type I error: we reject a hypothesis when it should be accepted.
Type II error: we accept a hypothesis when it should be rejected.

We will not place much emphasis on the Type II error in this book, but the reader should bear in mind that there is a chance of error whenever a null hypothesis is accepted.

In the next section the testing of a null hypothesis about the true proportion of successes in a binomial experiment (equivalently, the theoretical probability of a success on a single trial) will be treated more systematically, and examples will be given.

TEST OF $H_o: \pi = \pi_0$ VERSUS A SPECIFIED ALTERNATIVE

In the material that follows it is necessary to distinguish carefully between three symbols that will represent related quantities. The symbol π is familiar, of course, from the last chapter. It represents the theoretical probability of a success on a single binomial trial (or, equivalently, π represents the theoretical proportion of successes in a binomial experiment or population). In most situations the value of π is unknown. The symbol **p** will denote the random variable that is the observed proportion of successes. The random variable **p** takes on different values in different performances of the same binomial experiment. For instance, if the experiment of tossing a coin 100 times is performed five times the values of **p** might be 0·48, 0·43, 0·51, 0·49, and 0·54. A particular value of **p** is denoted by the symbol $\hat{p}$. For instance, in the first performance of the experiment of tossing a coin 100 times referred to above, $\hat{p} = 0·48$.

The most general hypothesis that is tested about the theoretical proportion of successes (equivalently, the probability of success on a single trial) is that π equals some specified number, say π_0. The most general alternative hypothesis is that π is not equal to the specified number. In this section we will first show the procedure for testing $H_o: \pi = \pi_0$ against $H_1: \pi \neq \pi_0$. Then we will discuss each of the alternative hypotheses $H_1: \pi > \pi_0$ and $H_1: \pi < \pi_0$.

It can be shown that if we are performing a binomial experiment for which the probability of success on a single trial is π_0, then the random variable **p**, the observed proportion of successes, has a distribution approximately normal, with mean π_0 and variance $\dfrac{\pi_0(1 - \pi_0)}{n}$. In order to standardize the random variable **p**, we subtract its mean and divide by its standard deviation. If the null hypothesis $H_o: \pi = \pi_0$ is true, then

$$z = \frac{\mathbf{p} - \pi_0}{\sqrt{\dfrac{\pi_0(1 - \pi_0)}{n}}}$$

is approximately standard normal. In any particular experiment the random variable **p** has the numerical value $\hat{p}$, and the random variable **z** has the numerical value

$$z = \frac{\hat{p} - \pi_0}{\sqrt{\dfrac{\pi_0(1 - \pi_0)}{n}}} \qquad (7.1)$$

This fact furnishes the basis for a test. If we are performing a binomial experiment for which the probability of success on a single trial is less than π_0, then we are likely to obtain a $\hat{p}$ value that is smaller than π_0. Thus, the numerator of the quantity in Formula (7.1) will be negative and we will obtain a negative z. If the observed $\hat{p}$ is considerably smaller than π_0 we will doubt that the hypothesis $H_0 : \pi = \pi_0$ is correct. We will interpret the phrase '$\hat{p}$ is considerably smaller than π_0' to mean that $\hat{p}$ is small enough to give us a value of z that is in the lower critical region of the test (see Fig. 7.1).

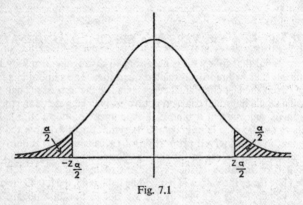

Fig. 7.1

The same line of reasoning applies to the other possibility. If we are performing a binomial experiment for which the probability of success on a single trial is greater than π_0, then the observed value of **p**, $\hat{p}$, will likely be larger than π_0. Thus, the numerator of the quantity in Formula (7.1) will be positive and we will obtain a positive value of z. If $\hat{p}$ is large enough to give us a value of z that is in the upper critical region of the test (also shown in Fig. 7.1) we reject the null hypothesis.

The critical region for the test of $H_0 : \pi = \pi_0$ versus $H_1 : \pi \neq \pi_0$ is composed of two pieces, one on each 'tail' of the normal distribution (which is approximating the binomial distribution whose mean and variance, according to H_0, are $n\pi_0$ and $n\pi_0(1 - \pi_0)$, respectively). The size of the critical region depends upon the value of α (size of the Type I error) that is chosen. Since the normal curve is symmetrical, and since we reject H_0 either for very small or for very large values of the observed proportion, it is reasonable to have the two parts of the critical region symmetrically located.

We are going to reject H_0 when it is true 100α per cent of the time (when $\alpha = 0.01$, we reject H_0 1 per cent of the time when H_0 is true; if $\alpha = 0.05$, we reject H_0 5 per cent of the time when H_0 is true, etc.). We want to reject half the time when $\hat{p}$ is large and half the time when $\hat{p}$ is small. So we select a value of z that is exceeded by 100 $\frac{\alpha}{2}$ per cent of the z values—a value of z to the right of which lies 100 . $\frac{\alpha}{2}$ per cent of the area under the standard normal curve. This z-value is denoted $z_{\alpha/2}$ in Fig. 7.1.

For example, $z_{0.10}$ is exceeded by 10 per cent of the z-values. Or, equivalently, 40 per cent of the area under the standard normal curve lies between 0 and $z_{0.10}$. Using notation from Chapter VI, we know that $A(z_{0.10}) = 0.40$; from the standard normal table we find that $A(1.28) = 0.39973$ and $A(1.29) = 0.40147$. We will not interpolate, but will simply select the value of z that gives us the value of $A(z)$ closest to 0.40. This value is 1.28; thus we have $z_{0.10} = 1.28$.

Because $\alpha = 0.05$ is frequently used, $z_{0.025}$ is also needed. We can find, as we did in the preceding paragraph, that the z-value to the right of which lies 2.5 per cent of the area under the curve is $z_{0.025} = 1.96$. Two other frequently used z-values, which the reader can verify, are $z_{0.05} = 1.645$ (this value is usually given to three decimal places, in contrast with the other values, which are only given to two decimal places) and $z_{0.005} = 2.58$.

We are now ready to state and to illustrate the procedure for testing the null hypothesis that π equals some specified number versus the alternative that it does not.

Procedure for Testing H_0: $\pi = \pi_0$ versus H_1: $\pi \neq \pi_0$

(1) Formulate the null and alternative hypotheses

$$H_0: \pi = \pi_0$$
$$H_1: \pi \neq \pi_0$$

(2) Decide upon a value for α. Note $z_{\alpha/2}$ and $-z_{\alpha/2}$, thereby determining the critical region.

(3) Select a random sample and compute $\hat{p}$ (or, for a 'textbook problem', compute $\hat{p}$ from the data given).

(4) Compute the quantity

$$z = \frac{\hat{p} - \pi_0}{\sqrt{\dfrac{\pi_0(1 - \pi_0)}{n}}}$$

(5) Compare the z found in (4) with $z_{\alpha/2}$ and $-z_{\alpha/2}$.

Reject H_0 if $z \geqslant z_{\alpha/2}$ or

if $z \leqslant -z_{\alpha/2}$.

Accept otherwise.

EXAMPLE: A man has just purchased a trick die which was advertised as not yielding the proper proportion of sixes. He wonders whether the advertising was correct, and would like to test the advertising claim by rolling the die 100 times. The 100 rolls yielded ten sixes. Should he conclude that the advertising was legitimate?

The 100 rolls of the die can be considered to be 100 trials composing a binomial experiment.

(1) If the advertising is false—if the die is just like any other die—then the proportion of successes (equivalently, the probability of success on any one trial) is equal to $\frac{1}{6}$. So the null hypothesis is $H_0: \pi = \frac{1}{6}$. The advertising did not state whether the

proportion of sixes should be greater or less than $\frac{1}{6}$. Therefore, we must use a two-sided alternative, $H_1: \pi \neq \frac{1}{6}$.

$$H_0: \pi = \tfrac{1}{6} = 0\cdot167$$
$$H_1: \pi \neq \tfrac{1}{6} = 0\cdot167$$

(2) Let $\alpha = 0\cdot05$. Then from the standard normal table we find $z_{0.025} = 1\cdot96$ and $-z_{0.025} = -1\cdot96$. The critical region (or rejection region) consists of all z-values to the right of (and including) $1\cdot96$ and to the left of (and including) $-1\cdot96$.

(3) The observed proportion is $\hat{p} = \frac{10}{100} = 0\cdot10$.

(4) Our test quantity is

$$z = \frac{0\cdot10 - 0\cdot167}{\sqrt{\dfrac{(0\cdot167)(0\cdot833)}{100}}} = \frac{0\cdot10 - 0\cdot167}{\sqrt{\dfrac{0\cdot139}{100}}} = \frac{-0\cdot067}{\sqrt{0\cdot00139}}$$

Then we have $\qquad z = \dfrac{-0\cdot067}{0\cdot03728} = -1\cdot80$

(5) $-1\cdot80 > -1\cdot96$

We accept H_0.

The experimental results do not furnish sufficient grounds upon which to reject H_0.

Procedure for Testing $H_0: \pi = \pi_0$ versus $H_1: \pi > \pi_0$

As stated previously, if H_0 is true, then the random variable

$$z = \frac{p - \pi_0}{\sqrt{\dfrac{\pi_0(1 - \pi_0)}{n}}}$$

is approximately standard normal. If $H_1: \pi > \pi_0$ is true, then we would expect our observed value, $\hat{p}$, to be larger than π_0, and z to be positive. The larger z is, the larger $\hat{p}$ is; and the larger $\hat{p}$ is, the more inclined we are to reject H_0: $\pi = \pi_0$ and accept $H_1: \pi > \pi_0$. (Values of $\hat{p}$ smaller than π_0 lead us to accept H_0, of course. It is understood that in accepting $H_0: \pi = \pi_0$, the fact that the true π might be less than π_0 is also being accepted. In fact, the null hypothesis is frequently stated to indicate that $H_0: \pi \leqslant \pi_0$. If a one-sided alternative is being tested the acceptance of the 'other side' is understood to be implicit in the acceptance of the null hypothesis, even when it is not explicitly stated.

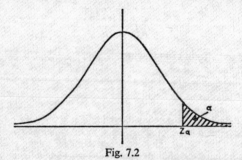

Fig. 7.2

For instance, if the null and alternative hypotheses are stated $H_0: \pi = 0\cdot6$ versus $H_1: \pi > 0\cdot6$, it is understood that H_0 could be written $H_0: \pi \leqslant 0\cdot6$—that by acceptance of $H_0: \pi = 0\cdot6$, we are implicitly accepting that π can be less than $0\cdot6$, also.) Therefore we will reject $H_0: \pi = \pi_0$ and accept H_1:

$\pi > \pi_0$ only for relatively large values of z. Hence we want the entire rejection region for the test to be on the right-hand tail of the normal distribution as in Fig. 7.2. For this reason tests of this sort are known as **one-tailed tests** or **one-sided tests**. The critical region depends upon α: the experimenter selects the size of the α, then finds the value z_α from the standard normal table; the critical region consists of all those z-values greater than or equal to z_α.

The step-by-step procedure for testing $H_o\colon \pi = \pi_0$ against $H_1\colon \pi > \pi_0$ is shown below:

(1) Formulate the null and alternative hypotheses

$$H_o\colon \pi = \pi_0$$
$$H_1\colon \pi > \pi_0$$

(2) Decide upon the α-level. Find z_α. Note the critical region.

(3) Draw a random sample (or perform a binomial experiment) and calculate the observed value of $\hat{p}$. (Or, for a 'textbook problem', compute $\hat{p}$ from the data given.)

(4) Compute the quantity $z = \dfrac{\hat{p} - \pi_0}{\sqrt{\dfrac{\pi_0(1 - \pi_0)}{n}}}$

(5) Compare the z found in (4) with z_α.
If $z \geqslant z_\alpha$, reject H_o.
Otherwise, accept H_o.

EXAMPLE: Mr. A and Mr. B are running for local public office in a large city. Mr. A says that only 30 per cent of the voters are in favour of a certain issue, a law to sell liquor on Sundays, say. Mr. B doubts A's statement and believes that more than 30 per cent favour such legislation. Mr. B pays for an independent organization to make a study of the situation. In a random sample of 400 voters, 160 favoured the legislation. What conclusions should the polling organization report to Mr. B?

(1) $H_o\colon \pi = 0.30,$
$H_1\colon \pi > 0.30.$

(2) We will select $\alpha = 0.05$. We have $z_{0.05} = 1.645$, and the critical region consists of all z-values greater than or equal to 1.645.

(3) From the data in the problem, $\hat{p} = \frac{160}{400} = 0.40$.

(4) $$z = \frac{0.40 - 0.30}{\sqrt{\dfrac{(0.30)(0.70)}{400}}} = \frac{0.10}{\dfrac{\sqrt{0.210}}{20}} = 4.36$$

(5) $$4.36 > 1.645$$

Reject H_o.

Procedure for Testing $H_o\colon \pi = \pi_0$ versus $H_1\colon \pi < \pi_0$

Once again, if H_o is true, then the random variable **p** has an approximately normal distribution with mean π_0 and variance $\dfrac{\pi_0(1 - \pi_0)}{n}$, and the random variable

$$z = \frac{\mathbf{p} - \pi_0}{\sqrt{\dfrac{\pi_0(1 - \pi_0)}{n}}}$$

is approximately standard normal.

If, for our sample, we obtain a $\hat{p}$ small enough to yield a z-value that is in the left-hand tail of the distribution (see Fig. 7.3), then we reject $H_o: \pi = \pi_0$ and accept $H_1: \pi < \pi_0$. Small values of $\hat{p}$ lead to small values of z, which lead us to accept the alternative hypothesis that the parameter π is less than π_0.

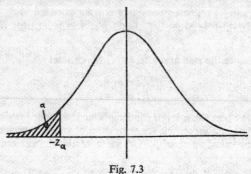

Fig. 7.3

The procedure is very similar to the preceding one-tailed test. In this case, however, we want all the rejection region on the left. If the size of our rejection region is α, then we need to find $-z_\alpha$. The rejection region is composed of all z-values less than $-z_\alpha$.

The step-by-step procedure is:

1. Formulate the null and alternative hypotheses.

$$H_o: \pi = \pi_0$$
$$H_1: \pi < \pi_0$$

2. Decide upon the α-level. Find $-z_\alpha$ from the standard normal tables. Note the critical region.

3. Draw a random sample, or perform a binomial experiment, and calculate the observed proportion, $\hat{p}$ (or compute $\hat{p}$ from the data given if a 'textbook problem' is being worked).

4. Compute the quantity $z = \dfrac{\hat{p} - \pi_0}{\sqrt{\dfrac{\pi_0(1 - \pi_0)}{n}}}$

5. Compare the z found in 4 with $-z_\alpha$. If $z \leqslant -z_\alpha$, reject H_o and accept H_1. Otherwise, accept H_o.

EXAMPLE: A coin was tossed 90 times, and 38 heads occurred. Test the hypothesis that the proportion of heads equals 0·5 versus the alternative that it is less.

(1) $$H_o: \pi = 0{\cdot}5$$
$$H_1: \pi < 0{\cdot}5$$

(2) Let $$\alpha = 0{\cdot}01$$
$$-z_{0.01} = -2{\cdot}33$$

The critical region is composed of all values of z less than or equal to $-2{\cdot}33$.

(3) $$\hat{p} = \tfrac{38}{90} = 0{\cdot}422$$

(4) $$z = \frac{0 \cdot 422 - 0 \cdot 500}{\sqrt{\dfrac{(0 \cdot 5)(0 \cdot 5)}{90}}} = \frac{-0 \cdot 078}{\sqrt{0 \cdot 00278}} = \frac{-0 \cdot 078}{0 \cdot 0527} = -1 \cdot 48$$

(5) $-1 \cdot 48 > -2 \cdot 33$. Accept H_0.

TESTS ABOUT THE MEAN OF A NORMAL DISTRIBUTION

In order to make a test about the mean of a normal distribution, we use the mean of a random sample drawn from a normal population: we use $\bar{x}$ (a particular numerical value of $\bar{x}$) to draw a conclusion about μ. We therefore need to know the distribution of $\bar{x}$.

An important theorem of mathematical statistics states that the mean of a random sample of size n drawn from a normal population with mean μ and variance σ^2 has a normal distribution with mean μ and variance σ^2/n. Using the notation introduced in Chapter VI, this can be stated more compactly:

$$\text{If } x = N(\mu, \sigma^2), \text{ then } \bar{x} = N\left(\mu, \frac{\sigma^2}{n}\right)$$

Recall that when we standardize a normal variable we obtain a standard normal variable. The standardized form of $\bar{x}$ is

$$z = \frac{\bar{x} - \mu}{\dfrac{\sigma}{\sqrt{n}}} \qquad (7.2)$$

which has a standard normal distribution. Symbolically, we have

$$\text{if } \bar{x} = N\left(\mu, \frac{\sigma^2}{n}\right) \quad \text{then} \quad z = \frac{\bar{x} - \mu}{\dfrac{\sigma}{\sqrt{n}}} = N(0,1)$$

Suppose we want to test the null hypothesis that μ equals some specified value, μ_0, against the alternative hypothesis that μ does not equal μ_0. Symbolically, we want to test $H_0: \mu = \mu_0$ versus $H_1: \mu \neq \mu_0$. If the null hypothesis is true—if the mean of the normal population from which we are sampling is truly μ_0—then the random variable $\dfrac{\bar{x} - \mu_0}{\dfrac{\sigma}{\sqrt{n}}}$ has a standard normal distribution.

The variance of the normal population, σ^2, must be known. If the variance is not known, we cannot use the standard normal distribution, and must use a distribution known as the t-distribution. The procedure is similar whether the variance is known or unknown. The case in which the variance is unknown is treated in the next chapter.

If we are sampling from a population whose mean is not μ_0, as specified, then the value of $\bar{x}$, $\bar{x}$, is likely to be either considerably larger than μ_0 or considerably smaller than μ_0, in either case leading us to reject $H_0: \mu = \mu_0$. If $H_0: \mu = \mu_0$ is true, then the quantity $\dfrac{\bar{x} - \mu_0}{\dfrac{\sigma}{\sqrt{n}}}$ is a value of a standard normal random variable, and this fact furnishes a convenient basis for the test of the null hypothesis.

The procedure for testing the null hypothesis that μ equals a specified value μ_0 against the alternative that μ is not equal to μ_0 is shown below. The procedure here is analogous to the one used in the previous section for testing a hypothesis about a proportion. The procedure for many succeeding tests is the same, too. In all these cases we compute a standardized quantity and compare it with values (which depend upon the value of α) from the appropriate table.

Procedure for Testing H_0: $\mu = \mu_0$ versus H_1: $\mu \neq \mu_0$

(1) Formulate the null and alternative hypotheses:

$$H_0: \mu = \mu_0$$
$$H_1: \mu \neq \mu_0$$

(2) Decide upon a value for α. Note $z_{\alpha/2}$ and $-z_{\alpha/2}$, and thereby determine the critical region of the test.

(3) Select a random sample from the normal population with known variance and compute $\bar{x}$ (or use the data given in the 'textbook problem' to compute $\bar{x}$).

(4) Compute the quantity $z = \dfrac{\bar{x} - \mu_0}{\dfrac{\sigma}{\sqrt{n}}}$

(5) Compare the value of z found in (4) with $z_{\alpha/2}$ and $-z_{\alpha/2}$.

Reject H_0 if $z \geqslant z_{\alpha/2}$ or if $z \leqslant -z_{\alpha/2}$.
Accept otherwise.

EXAMPLE: Past experience has shown that the scores of students who take a certain mathematics test are normally distributed with mean 75 and variance 36.

The Mathematics Department members would like to know whether this year's group of 16 students is typical. They decide to test the hypothesis that this year's students are typical versus the alternative that they are not typical. When the students take the test the average score is 82. What conclusion should be drawn?

Following the steps in the above procedure we have the following:

(1) $$H_0: \mu = 75,$$
$$H_1: \mu \neq 75$$

Note that the null and alternative hypotheses are formulated before the data are gathered and the $\bar{x}$ computed. The sample data should not influence the choice of μ_0, or the choice of what alternative hypothesis to use.

(2) Let $$\alpha = 0 \cdot 10$$
$$z_{0.05} = 1 \cdot 645$$
$$-z_{0.05} = -1 \cdot 645$$

The choice of a value for α is completely at the discretion of the experimenter. We could have chosen any other value (usually a small value; there is not much sense in testing a null hypothesis if it is to be rejected a very large proportion of the time, more than 20 per cent of the time, say).

(3) $$\bar{x} = 82$$

The problem as stated gives us the value of $\bar{x}$. We must assume that the sample is random.

(4) In this problem $\sigma = \sqrt{36} = 6$, $n = 16$, and $\mu_0 = 75$. Therefore, we have

$$z = \frac{82 - 75}{6/4} = 4\cdot67$$

(5) $$4\cdot67 > 1\cdot645$$

We reject H_0: $\mu = 75$ and accept H_1: $\mu \neq 75$. On the basis of the evidence, the null hypothesis is unlikely to be correct. We conclude that this group of students is not typical; it seems apparent that it is superior.

Procedure for Testing H_0: $\mu = \mu_0$ versus H_1: $\mu > \mu_0$

(1) Formulate the null and alternative hypotheses.

$$H_0: \mu = \mu_0$$
$$H_1: \mu > \mu_0$$

(2) Select a value for α. Note z_α from the standard normal table. The critical region lies to the right of z_α.

(3) Select a random sample and compute $\bar{x}$ (or compute $\bar{x}$ from the data given in the 'textbook problem').

(4) Compute the quantity $z = \dfrac{\bar{x} - \mu_0}{\dfrac{\sigma}{\sqrt{n}}}$

(5) Compare z with z_α. Reject H_0 if $z \geqslant z_\alpha$. Accept otherwise.

EXAMPLE: Given that a random sample of size 9 yielded $\bar{x} = 23$, test H_0: $\mu = 21$ versus H_1: $\mu > 21$. Let $\alpha = 0\cdot01$.
Given $\sigma = 4$.

(1) $$H_0: \mu = 21$$
$$H_1: \mu > 21$$

(2) $$\alpha = 0\cdot01$$
$$z_{0.01} = 2\cdot33$$

(3) $$\bar{x} = 23$$

(4) $$z = \frac{23 - 21}{4/3} = 1\cdot5$$

(5) $$1\cdot5 < 2\cdot33$$

Accept H_0: $\mu = 21$.

Procedure for testing H_0: $\mu = \mu_0$ versus H_1: $\mu < \mu_0$

The procedure is exactly the same as the previous one except that the alternative hypothesis is different and that our critical region is on the left instead of on the right. With the necessary modification, steps (1), (2), and (5) become:

(1) Formulate the null and alternative hypotheses.

$$H_0: \mu = \mu_0$$
$$H_1: \mu < \mu_0$$

(2) Select a value for α.
Note $-z_\alpha$ from the standard normal tables. The critical region lies to the left of $-z_\alpha$.

(5) Compare z with $-z_\alpha$. If $z \leqslant -z_\alpha$, reject H_0. Accept H_0 otherwise. Steps (3) and (4) are the same.

EXERCISES

1. Test the null hypothesis, H_o: $\pi = 0.6$, against the alternative hypothesis, H_1: $\pi > 0.6$, if a sample of size 150 from a binomial population gave 100 successes. Let $\alpha = 0.05$.

2. If 100 tosses of a coin gave 63 heads, would you conclude that the coin is biased in favour of heads? Let $\alpha = 0.01$.

3. A sample of size 78 from a binomial population gave 35 successes. Test the null hypothesis that the true proportion of successes is 0.55 against the alternative that it is less.

4. A sample of size 52 from a binomial population gave 15 successes. Test H_o: $\pi = 0.4$ versus H_1: $\pi < 0.4$.

5. Suppose that in order to test the hypothesis that $\pi = 0.6$ against the alternative that $\pi < 0.6$, we decide to obtain a sample of size 100 and reject H_o if we obtain fewer than 48 successes. What is the approximate size of the Type I error? If the value of π is really 0.5, what is the size of the Type II error?

6. A sample of size 16 from a normal population with known variance 256 gave $\bar{x} = 40$. Test H_o: $\mu = 45$ versus H_1: $\mu \neq 45$. Let $\alpha = 0.01$.

7. Test the hypothesis that the mean of a normal population with known variance 70 is 31, if a sample of size 13 gave $\bar{x} = 34$. Let the alternative hypothesis be H_1: $\mu > 31$, and let $\alpha = 0.10$.

CHAPTER VIII

MORE TESTS OF HYPOTHESES

INTRODUCTION

The important ideas involved in the testing of a statistical hypothesis were discussed in the first two sections of Chapter VII. In the remainder of that chapter tests of hypotheses about the theoretical proportion of successes in a binomial population and about the theoretical mean of a normal population were discussed and illustrated. Recall that in order to test a hypothesis about the mean of a population, it had to be assumed that the population was normal and that its variance was known.

In this chapter we will learn how to test hypotheses about the mean of a normal population when the population variance is not known and to test hypotheses about the mean of populations that are not normal. The procedure followed to test hypotheses in both of these situations is similar to that followed in the last section in Chapter VII.

After treating these topics we will turn to situations that are more complicated. Up to this point we have been concerned with testing a hypothesis about a parameter (either π or μ) from a single population. Frequently, however, it is necessary to compare two populations. We compare two binomial populations by testing hypotheses about the difference of the two proportions of successes. Two populations (which are not binomial ones) are compared by testing a hypothesis about the difference of the two population means. There are several subcases here, depending upon whether the two populations are normal or not, and upon whether the variances are known or unknown.

All of the tests that were discussed in Chapter VII or mentioned above are about either means or proportions. Tests of hypotheses about the variance of a population are discussed in the section titled 'Tests about the Variance of a Normal Population'. When two populations are being examined we may want to test whether the variances of the two populations are equal. This is covered in the section titled 'Tests about the Ratio of Two Variances'.

TEST OF H_o: $\mu = \mu_0$, NORMAL POPULATION, σ^2 UNKNOWN

In Chapter VII the test of the null hypothesis H_o: $\mu = \mu_0$ versus a specified alternative was treated. It was stated that if a random sample of size n is drawn from a normal population with unknown mean μ and known variance σ^2 the sample mean $\bar{x}$ will have a normal distribution with unknown mean μ and known variance σ^2/n.

It is only rarely, however, that the variance of the population from which we are sampling is known. Usually the variance as well as the mean is unknown, and therefore the theory presented in Chapter VII is not applicable. It can be shown that if the sample mean $\bar{x}$ is based on a random sample of size n selected from a normal population with unknown mean μ and unknown variance σ^2, then the random variable

$$\frac{\bar{x} - \mu_0}{s/\sqrt{n}}$$

has a distribution known as the **t-distribution**. This is also known as 'student's' t-distribution after its discoverer Gosset, who published under the pseudonym of 'Student'. The t-distribution depends upon a single parameter, known as the **number of degrees of freedom**. In the present situation the number of degrees of freedom is one less than the sample size, namely $n - 1$. The graph of the t-distribution resembles the graph of the standard normal distribution: they are both symmetric, bell-shaped curves with mean equal to zero. The graph of the t-distribution, however, is lower at the centre and higher at the extremities than the standard normal curve. These differences are more pronounced when the number of degrees of freedom is small; as the number of degrees of freedom increases, the graph of the t-distribution resembles the standard normal curve more and more. For 'infinite' degrees of freedom (more than 120, the largest value in the table) percentage points of the t-distribution are equal to those of the normal distribution.

In Fig. 8.1 the graph of a t-distribution with 3 degrees of freedom is superimposed on the graph of the standard normal distribution. The similarities and differences remarked upon above should be noted. The number of degrees of freedom, 3, is very small, hence the differences are very pronounced, relatively speaking. For a moderately large number of degrees of freedom the t-distribution is practically indistinguishable from the standard normal distribution.

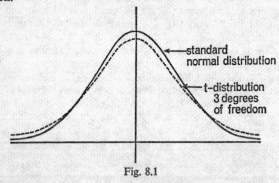

standard normal distribution

t-distribution 3 degrees of freedom

Fig. 8.1

For a t-distribution with $n - 1$ degrees of freedom the symbol $t_\alpha(n - 1)$ denotes the t-value to the right of which lies 100α per cent of the area under the graph of that distribution: $t_\alpha(n - 1)$ is the upper α-point of the t-distribution with $n - 1$ degrees of freedom (see Fig. 8.2). For instance, $t_{0.05}(10)$ represents the t-value to the right of which lies 5 per cent of the area under the curve of the t-distribution that has 10 degrees of freedom.

The t-table in the Appendix is arranged to give the value $t_\alpha(n - 1)$ for several frequently used values of α and for a number of values of $n - 1$ from 1 to 120. The α-values in our table are 0·005, 0·01, 0·025, 0·05, and 0·10. In order to find a t-value for a particular α-value and a particular number of degrees of freedom, the table is entered under the column which corresponds to the desired α-value. Then read down that particular column until you find the required *t*-value on the line opposite the appropriate number of degrees of freedom.

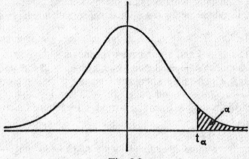

Fig. 8.2

As an example, we will find $t_{0.05}(13)$. In the 0·05 column, opposite 13 degrees of freedom, we read 1·771. To find $t_{0.01}(24)$ we enter the 0·01 column, and opposite 24 degrees of freedom we read 2·492.

Since the *t*-distribution is symmetrical about the value $t = 0$, the lower α-points can be obtained from the upper α-points. Therefore the *t*-table need contain only the upper α-points. The relationship between the lower and upper points is

$$t_{1-\alpha} = -t_\alpha \tag{8.1}$$

in words, the lower α-point is the negative of the upper α-point (see Fig. 8.3).

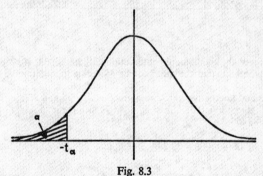

Fig. 8.3

EXAMPLE: Find $t_{0.975}(5)$.
The relation in Formula (8.1) tells us that

$$t_{0.975} = -t_{0.025}$$

In the *t*-table we find $t_{0.025}(5) = 2·57$; therefore, $t_{0.975}(5) = -2·57$.

If we wish to find the *t*-value that corresponds to a number of degrees of freedom which is not given in the table we will use the nearest degrees of freedom that are in the table.

The procedure for testing the null hypothesis H_o: $\mu = \mu_0$ versus whatever alternative might be specified is the same as in Chapter VII. First, the experimenter decides what the null and alternative hypotheses are to be and then selects an α-level for the test. After these preliminaries the random sample is drawn, the test statistic calculated, and the hypothesis accepted or rejected. These steps are listed below for the alternative hypothesis H_1: $\mu \neq \mu_0$.

Procedure for Testing H_o: $\mu = \mu_0$ versus H_1: $\mu \neq \mu_0$ (Variance Unknown)

(1) Formulate the null and alternative hypotheses.

(2) Decide upon the α-level. Look up $t_{\alpha/2}(n-1)$ in the table. We are testing H_o against a two-tailed alternative, so the critical region consists of two pieces: $t \leqslant -t_{\alpha/2}(n-1)$ and $t \geqslant t_{\alpha/2}(n-1)$ (see Fig. 8.4).

(3) Draw a random sample from the normal population. Calculate $\bar{x}$ and s.

(4) Calculate $t = \dfrac{\bar{x} - \mu_0}{s/\sqrt{n}}$

(5) Compare the t-value in step (4) with $t_{\alpha/2}(n-1)$ and $-t_{\alpha/2}(n-1)$. Reject H_o if $t \leqslant -t_{\alpha/2}(n-1)$ or if $t \geqslant t_{\alpha/2}(n-1)$.

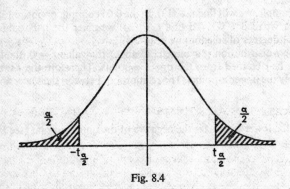

Fig. 8.4

EXAMPLE: Given the eight sample observations 31, 29, 26, 33, 40, 28, 30, and 25, test the null hypothesis that the mean equals 35 versus the alternative that it does not. Let $\alpha = 0.01$.

(1) The null and alternative hypotheses are H_o: $\mu = 35$ and H_1: $\mu \neq 35$.

(2) We are given $\alpha = 0.01$. This is a two-tailed test: $t_{0.005}(7) = 3.499$ and $-t_{0.005}(7) = -3.499$.

$$(3) \qquad \bar{x} = \frac{\sum x_i}{8} = 30.25$$

$$s^2 = \frac{1}{7}\left[\sum x_i^2 - \frac{(\sum x_i)^2}{8}\right]$$

$$= \frac{1}{7}\left[7\,476 - \frac{(242)^2}{8}\right] = \frac{155.5}{7} = 22.21$$

$$s = \sqrt{22.2} = 4.71$$

$$(4) \qquad t = \frac{30.25 - 35}{4.71/\sqrt{8}} = \frac{-4.75\sqrt{8}}{4.71} = -2.85$$

(5) $-2.85 > -3.499$. Accept H_0.

Note that if we had let $\alpha = 0.05$ the null hypothesis would have been rejected.

Procedure for Testing H_0: $\mu = \mu_0$ *versus* H_1: $\mu < \mu_0$ (*Variance Unknown*)

(1) Formulate the null and alternative hypotheses, H_0: $\mu = \mu_0$ versus H_1: $\mu < \mu_0$.

(2) Decide upon the α-level, and look up $t_\alpha(n - 1)$. The critical region consists of all *t*-values that are less than or equal to $-t_\alpha(n - 1)$ (see Fig. 8.3).

(3) Draw a random sample from the normal population. Calculate $\bar{x}$ and s.

(4) Calculate $t = \dfrac{\bar{x} - \mu_0}{s/\sqrt{n}}$

(5) Reject H_0 if $t \leqslant -t_\alpha(n - 1)$.

Procedure for testing H_0: $\mu = \mu_0$ *versus* H_1: $\mu > \mu_0$ (*Variance Unknown*)

(1) Formulate the null and alternative hypotheses, H_0: $\mu = \mu_0$ and H_1: $\mu > \mu_0$.

(2) Decide upon the α-level, and look up $t_\alpha(n - 1)$. The critical region consists of all *t*-values that are greater than or equal to $t_\alpha(n - 1)$ (see Fig. 8.2).

(3) Draw a random sample from the normal population. Calculate $\bar{x}$ and s.

(4) Calculate $t = \dfrac{\bar{x} - \mu_0}{s/\sqrt{n}}$

(5) Reject H_0 if $t \geqslant t_\alpha(n - 1)$.

The assumption that the population from which the sample is drawn is a normal population cannot be relaxed if the sample is small. If the sample is large—more than thirty, say—then the population does not need to be assumed to be a normal one, even when the variance is unknown.

TESTS ABOUT THE MEAN OF A NON-NORMAL POPULATION

Statistical tests of hypotheses about the mean of a distribution that is not normal, and whose variance may be unknown, are only approximate tests. One of the most important theorems of statistics, known as the **Central Limit Theorem**, states that if a random sample of size n is drawn from a population with mean μ and variance σ^2 (there is one other condition, but it need not concern us), then as n increases, the distribution of $\dfrac{\bar{x} - \mu}{\sigma/\sqrt{n}}$ approaches the standard normal distribution.

This is a remarkable theorem. We can be sampling from any population whatsoever (we must assume that the population has a finite variance, but the reader need not worry about this), and for a large sample $\dfrac{\bar{x} - \mu}{\sigma/\sqrt{n}}$ is approximately standard normal. We will understand 'a large sample' to mean that the sample contains thirty or more observations. The larger the sample is, the closer the distribution of $\dfrac{\bar{x} - \mu}{\sigma/\sqrt{n}}$ is to a standard normal distribution. Although in the statement of the theorem the population standard deviation is presumed known, σ can be replaced by the sample standard deviation s without any serious error when a large sample is available. Then we can say that, for samples sufficiently large (about thirty or more), the quantity $\dfrac{\bar{x} - \mu}{s/\sqrt{n}}$ is also approximately normal, with larger samples giving better approximations.

The preceding discussion furnishes the basis for a test of the null hypothesis H_o: $\mu = \mu_0$ versus a specified alternative: the quantity $\dfrac{\bar{x} - \mu}{\sigma/\sqrt{n}}$ or $\dfrac{\bar{x} - \mu}{s/\sqrt{n}}$ is calculated and referred to normal tables.

This is the third section which has been devoted to the testing of hypotheses about population means, and care should be taken to distinguish among the three situations:

(1) The population is normal and the variance is known. Then the random variable $\dfrac{\bar{x} - \mu}{\sigma/\sqrt{n}}$ is *exactly* standard normal, no matter how small the sample size n is.

(2) The population is normal and the variance is unknown. The random variable $\dfrac{\bar{x} - \mu}{s/\sqrt{n}}$ has *exactly* a t-distribution with $n - 1$ degrees of freedom, no matter how small the sample size.

(3) The population is not normal and the variance may or may not be known. Then the random variable $\dfrac{\bar{x} - \mu}{\sigma/\sqrt{n}}$ or the random variable $\dfrac{\bar{x} - \mu}{s/\sqrt{n}}$ (the one used depending on whether the variance is known or unknown) is *approximately* normal if the sample size is sufficiently large (at least thirty). This is the situation being treated in the present section.

We must have a large sample if the population from which we are sampling is not normal. If the population is normal, then a small sample is sufficient, and we use either the normal table or the t-table to test an hypothesis about the mean of the population, depending upon whether the population variance is known or unknown.

Procedure for Testing H_o: $\mu = \mu_0$ versus a Specified Alternative, Large Sample

(1) Formulate the null and alternative hypotheses.
(2) Decide upon the α-level of the test.

(a) H_1: $\mu \neq \mu_0$. This is a two-tailed alternative. Look up $z_{\alpha/2}$. The critical region consists of two pieces: $z \leqslant -z_{\alpha/2}$ and $z \geqslant z_{\alpha/2}$ (see Fig. 7.1).
(b) H_1: $\mu > \mu_0$. This is a one-tailed test. Look up z_α. The critical region consists of all values greater than or equal to z_α (see Fig. 7.2).
(c) H_1: $\mu < \mu_0$. This is a one-tailed test. Look up z_α. The critical region consists of all values less than or equal to $-z_\alpha$ (see Fig. 7.3).

(3) Draw the random sample. Calculate $\bar{x}$. Calculate s if σ is unknown.
(4) Calculate $z = \dfrac{\bar{x} - \mu_0}{\sigma/\sqrt{n}}$ or $z = \dfrac{\bar{x} - \mu_0}{s/\sqrt{n}}$, whichever is appropriate.
(5) Reject H_o if z is in the critical region. Accept H_o otherwise.

EXAMPLE: A sample of size 49 yielded the values $\bar{x} = 87 \cdot 3$ and $s^2 = 162$. Test the hypothesis that $\mu = 95$ versus the alternative that it is less. Let $\alpha = 0 \cdot 01$.
(1) We are testing H_o: $\mu = 95$ versus H_1: $\mu < 95$.
(2) $\alpha = 0 \cdot 01$. We find $-z_{0 \cdot 01} = -2 \cdot 33$. The critical region consists of all values of z less than or equal to $-2 \cdot 33$.
(3) We are given $\bar{x} = 87 \cdot 3$ and $s^2 = 162$; $s = \sqrt{162} = 12 \cdot 73$.

(4) $$z = \frac{87 \cdot 3 - 95}{12 \cdot 73 / \sqrt{49}} = \frac{-7 \cdot 7}{12 \cdot 73 / 7} = \frac{-7 \cdot 7}{1 \cdot 82} = -4 \cdot 23$$

(5) $-4 \cdot 23 < -2 \cdot 33$. Reject H_0.

EXAMPLE: The mean score on a widely given history examination is 75. A history teacher at a very large university wants to determine whether there is statistical evidence for claiming that this year's class is not average. Given the following scores, and assuming that the students in his class are a random sample from the population of students at the university, test the appropriate hypothesis versus the appropriate alternative.

The test scores are:

94	69	89	49	88	89	85
95	55	93	86	62	83	96
48	51	69	74	83	71	89
58	89	81	79	52	73	
75	91	68	100	63	81	

(1) We are testing H_0: $\mu = 75$ versus H_1: $\mu \neq 75$. A two-tailed alternative is selected because there is no prior knowledge of whether the present class is better or worse than average.

(2) Let $\alpha = 0 \cdot 05$. (This is arbitrarily chosen by the experimenter.) $z_{0.025} = 1 \cdot 96$ and $-z_{0.025} = -1 \cdot 96$. The critical region consists of two pieces: $z \leqslant -1 \cdot 96$ and $z \geqslant 1 \cdot 96$.

(3) $$\sum x = 2528, \sum x^2 = 201\,026, n = 33$$
$$\bar{x} = \frac{\sum x}{n} = \frac{2528}{33} = 76 \cdot 6$$
$$s^2 = \frac{1}{32} \left[201\,026 - \frac{(2528)^2}{33} \right]$$
$$= \frac{1}{32} [201\,026 - 193\,660]$$
$$= \frac{1}{32} (7366) = 230 \cdot 2$$
$$s = 15 \cdot 17$$

(4) $$z = \frac{76 \cdot 6 - 75}{15 \cdot 17 / \sqrt{33}} = \frac{1 \cdot 6}{2 \cdot 64} = 0 \cdot 61$$

(5) $0 \cdot 61 < 1 \cdot 96$. Accept H_0.

The instructor concludes that there is no statistical basis for concluding that the class this year is not average.

TESTS ABOUT THE DIFFERENCE OF TWO PROPORTIONS

The testing of hypotheses about the theoretical proportion of successes in a binomial experiment was treated in Chapter VII. Occasionally we may have two binomial populations and we want to test whether the proportion of successes is the same for both or, more generally, that the difference between the two proportions of successes is a specified number.

The true proportion of successes in Population 1 will be denoted by π_1, and the true proportion in Population 2 will be denoted by π_2. Do not confuse the symbol π, which is used here to denote a parameter of a binomial population, with the use of the same symbol to denote the transcendental number that is approximately equal to $3 \cdot 1416$.

The term **estimator** refers to a random variable, particular numerical values of which are used as **estimates** of the true (and unknown) value of some parameter. An estimator is a random variable; an estimate is a particular numerical value of an estimator for a particular random sample.

The random variable p_1 (the sample proportion of successes) is used as an estimator of π_1. The variable p_1 is a random variable because in different samples (of size n_1, say) it takes on different values with certain probabilities, and it cannot be predicted in advance what value p_1 will have in a particular sample. The random variable p_1, based on a random sample of size n_1 from Population 1, has a distribution with theoretical mean π_1 and theoretical variance $\dfrac{\pi_1(1 - \pi_1)}{n_1}$. The value that the random variable p_1 takes on in a particular sample will be denoted by $\hat{p}_1$. The observed value of the random variable in a particular sample is the observed number of successes, x_1, divided by n_1, the size of the sample:

$$\hat{p}_1 = x_1/n_1$$

Similarly, the random variable p_2, based on a sample of size n_2, is used as an estimator for π_2, and has a distribution with theoretical mean π_2 and theoretical variance $\dfrac{\pi_2(1 - \pi_2)}{n_2}$. The value that the random variable p_2 takes on in a particular sample will be denoted by $\hat{p}_2$. The observed value of p_2 is thus

$$\hat{p}_2 = x_2/n_2$$

where
$x_2 =$ the number of successes in the sample; and
$n_2 =$ the size of the sample.

The most general hypothesis that we can test about the difference of two theoretical proportions π_1 and π_2 is that it equals some specified value. The traditional symbol for 'difference' is Δ (delta, the fourth letter of the Greek alphabet). Since the difference is specified by a null hypothesis, we will attach a zero subscript. Thus, the difference between the theoretical proportions specified by the null hypothesis is denoted Δ_0, and the null hypothesis itself is written $H_0 : \pi_1 - \pi_2 = \Delta_0$.

Usually, we are interested in testing that the theoretical proportion of successes is the same in one population as in the other. This is equivalent to testing that the difference between the two proportions is zero (that Δ_0 equals zero); that is, we usually test $H_0 : \pi_1 - \pi_2 = 0$. This null hypothesis can be written $H_0 : \pi_1 = \pi_2$, of course, but we will follow the same pattern as in the general statement $H_0 : \pi_1 - \pi_2 = \Delta_0 \neq 0$.

The following statement is a very important theorem, although we need not be concerned with its proof in this book: if x and y are two independent random variables with variances σ_x^2 and σ_y^2, then the variance of $x - y$ is $\sigma_x^2 + \sigma_y^2$. In other words, the variance of the difference of two independent random variables equals the sum of the separate variances. This theorem is needed in the discussion below, as well as in the next section.

The variance of the random variable p_1 is $\pi_1(1 - \pi_1)/n_1$ (stated earlier in this section). Also, the variance of the random variable p_2 is $\pi_2(1 - \pi_2)/n_2$. By the above theorem, therefore, the variance of the random variable $p_1 - p_2$ is $\pi_1(1 - \pi_1)/n_1 + \pi_2(1 - \pi_2)/n_2$.

It can be shown that the standardized random variable

$$z = \frac{(p_1 - p_2) - (\pi_1 - \pi_2)}{\sqrt{\dfrac{\pi_1(1 - \pi_1)}{n_1} + \dfrac{\pi_2(1 - \pi_2)}{n_2}}} \tag{8.2}$$

has a distribution that is approximately standard normal.

If $H_o: \pi_1 - \pi_2 = \Delta_0$ is true, then the random variable

$$z = \frac{(p_1 - p_2) - \Delta_0}{\sqrt{\dfrac{\pi_1(1 - \pi_1)}{n_1} + \dfrac{\pi_2(1 - \pi_2)}{n_2}}} \tag{8.3}$$

has a distribution that is approximately standard normal. This fact furnishes the basis for a test of $H_o: \pi_1 - \pi_2 = \Delta_0$.

In order to use the value of the standardized random variable z to test a hypothesis about $\pi_1 - \pi_2$, the variance of $p_1 - p_2$, namely

$$\frac{\pi_1(1 - \pi_1)}{n_1} + \frac{\pi_2(1 - \pi_2)}{n_2}$$

must be known or estimated. The variance is not known, because the values of the parameters π_1 and π_2 are not known. But it is natural to estimate these theoretical proportions by the observed sample proportions $\hat{p}_1$ and $\hat{p}_2$. Therefore, the theoretical variance of the random variable $p_1 - p_2$ is estimated by the quantity

$$\frac{\hat{p}_1(1 - \hat{p}_1)}{n_1} + \frac{\hat{p}_2(1 - \hat{p}_2)}{n_2}$$

If we let $1 - \hat{p}_1 = \hat{q}_1$ and $1 - \hat{p}_2 = \hat{q}_2$, this estimated variance can be written more compactly as

$$\frac{\hat{p}_1\hat{q}_1}{n_1} + \frac{\hat{p}_2\hat{q}_2}{n_2}$$

For any particular experiment the observed value of the random variable $p_1 - p_2$ is $\hat{p}_1 - \hat{p}_2$, and the observed value of the random variable z in Formula (8.3) is

$$z = \frac{(\hat{p}_1 - \hat{p}_2) - \Delta_0}{\sqrt{\dfrac{\hat{p}_1\hat{q}_1}{n_1} + \dfrac{\hat{p}_2\hat{q}_2}{n_2}}} \tag{8.4}$$

We know that if H_o is true the random variable in Formula (8.3) is approximately standard normal. Therefore we can test $H_o: \pi_1 - \pi_2 = \Delta_0$ by noting whether the numerical value of z in (8.4) is in the critical region of the test. The step-by-step procedure for testing $H_o: \pi_1 - \pi_2 = \Delta_0$ is given below.

Procedure for Testing $H_o: \pi_1 - \pi_2 = \Delta_0$ (Δ_0 is not zero)

(1) Formulate the null and alternative hypotheses.

(2) Decide upon the α-level of the test.

  (a) If the alternative hypothesis is $H_1: \pi_1 - \pi_2 \neq \Delta_0$, find $z_{\alpha/2}$ and $-z_{\alpha/2}$. The critical region of the test consists of two parts: $z \leqslant -z_{\alpha/2}$ and $z \geqslant z_{\alpha/2}$ (see Fig. 7.1).

(b) If the alternative hypothesis is $H_1: \pi_1 - \pi_2 < \Delta_0$, find $-z_\alpha$. The critical region of the test consists of all values of z less than or equal to $-z_\alpha$.

(c) If the alternative hypothesis is $H_1: \pi_1 - \pi_2 > \Delta_0$, find z_α. The critical region of the test consists of all values of z greater than or equal to z_α.

(3) Perform the experiment or draw a random sample from each of the binomial populations. Calculate $\hat{p}_1$ and $\hat{p}_2$ (three decimal place accuracy is recommended).

(4) Calculate $z = \dfrac{(\hat{p}_1 - \hat{p}_2) - \Delta_0}{\sqrt{\dfrac{\hat{p}_1\hat{q}_1}{n_1} + \dfrac{\hat{p}_2\hat{q}_2}{n_2}}}$

(5) Reject H_o if z is in the critical region.

Example: The Service chief of a certain country suspects that the proportion of men from urban areas who are physically unfit for military service is more than 5 percentage points greater than the proportion of physically unfit men from rural areas. He decides to treat the men called for a physical examination from urban areas during the next month as a random sample from a binomial population, and those from the rural areas as a random sample from a second binomial population. During the next month 3,214 men were called from urban areas and 2,011 from rural areas. There were 1,078 physical rejects from the urban areas and 543 from rural areas.

Formulate the appropriate null and alternative hypotheses, and test the null hypothesis at the $\alpha = 0.05$ level.

(1) Let $\pi_1 =$ the true proportion of physically unfit from urban areas.

Let $\pi_2 =$ the true proportion of physically unfit from rural areas.

The null and alternative hypotheses are

$$H_0: \pi_1 - \pi_2 = 0.05 \quad \text{and}$$
$$H_1: \pi_1 - \pi_2 > 0.05$$

(2) We are given $\alpha = 0.05$; $z_{0.05} = 1.645$. The critical region consists of all z-values greater than or equal to 1.645.

(3)
$$x_1 = 1078, \ n_1 = 3214$$
$$p_1 = \frac{1078}{3214} = 0.335$$
$$x_2 = 543, \ n_2 = 2011$$
$$p_2 = \frac{543}{2011} = 0.270$$

(4)
$$z = \frac{(0.335 - 0.270) - 0.05}{\sqrt{\dfrac{(0.335)(0.665)}{3214} + \dfrac{(0.270)(0.730)}{2011}}}$$

$$= \frac{0.015}{\sqrt{0.0000693 + 0.0000980}}$$

$$= \frac{0.015}{\sqrt{0.0001673}} = \frac{0.015}{0.0129} = 1.16$$

(5) $1.16 < 1.645$. Accept H_o.

The procedure is somewhat different when we are testing the hypothesis that there is no difference between the proportions of successes in two binomial populations, $H_o: \pi_1 - \pi_2 = 0$. (This is equivalent to testing that the true proportion of successes in each population is equal to some value, say π, H_o: $\pi_1 = \pi_2 = \pi$.) Under the assumption that H_o is true, the random variable

$$z = \frac{(p_1 - p_2) - 0}{\sqrt{\dfrac{\pi(1 - \pi)}{n_1} + \dfrac{\pi(1 - \pi)}{n_2}}} \qquad (8.5)$$

is approximately standard normal.

As usual, the true value of the standard deviation of the random variable $p_1 - p_2$ is unknown, because the value of π is unknown. Since π_1 and π_2 are both assumed to be equal to π, we combine the numbers of successes from each population and combine the sample sizes from each population to get what is known as a **pooled estimate** of π, denoted $\hat{p}$: that is,

$$\hat{p} = \frac{\text{Total number of successes}}{\text{Total number of trials}}$$

If we denote the number of successes in the first sample by x_1, and the number of successes in the second sample by x_2, then $\hat{p}$ is given by the formula

$$\hat{p} = \frac{x_1 + x_2}{n_1 + n_2} \qquad (8.6)$$

which is equivalent to the formula

$$\hat{p} = \frac{n_1 \hat{p}_1 + n_2 \hat{p}_2}{n_1 + n_2} \qquad (8.7)$$

Under the assumption that the null hypothesis $H_o: \pi_1 - \pi_2 = 0$, the random variable

$$z = \frac{(p_1 - p_2) - (0)}{\sqrt{\dfrac{\hat{p}(1 - \hat{p})}{n_1} + \dfrac{\hat{p}(1 - \hat{p})}{n_2}}} \qquad (8.8)$$

has a distribution that is approximately standard normal.

The numerical value of Formula (8.8) that is obtained in a particular experiment,

$$z = \frac{(\hat{p}_1 - \hat{p}_2) - (0)}{\sqrt{\dfrac{\hat{p}(1 - \hat{p})}{n_1} + \dfrac{\hat{p}(1 - \hat{p})}{n_2}}} \qquad (8.9)$$

is used to test H_o. Note that the individual sample proportions are used in the numerator, whereas the pooled proportion is used in the denominator.

Procedure for Testing $H_o: \pi_1 - \pi_2 = 0$

(1) Formulate the null and alternative hypotheses.
(2) Decide upon the α-level of the test.

    (a) If the alternative hypothesis is $H_1: \pi_1 - \pi_2 \neq \Delta_0$, find $z_{\alpha/2}$ and $-z_{\alpha/2}$. The critical region of the test consists of all z-values less than or equal to $-z_{\alpha/2}$ or greater than or equal to $z_{\alpha/2}$ (see Fig. 7.1).
    (b) If the alternative hypothesis is $H_1: \pi_1 - \pi_2 < \Delta_0$, find $-z_\alpha$. The

critical region of the test consists of all values of z less than or equal to $-z_\alpha$ (see Fig. 7.3).

(c) If the alternative hypothesis is $H_1: \pi_1 - \pi_2 > \Delta_0$, find z_α. The critical region of the test consists of all values of z greater than or equal to z_α (see Fig. 7.2).

(3) Obtain a random sample from each of two binomial populations.

(4) Calculate $\hat{p}_1 = \dfrac{x_1}{n_1}$ and $\hat{p}_2 = \dfrac{x_2}{n_2}$

Calculate $\hat{p} = \dfrac{x_1 + x_2}{n_1 + n_2}$

(5) Calculate z from Formula (8.9).

(6) Reject H_o if z is in the critical region.

EXAMPLE. Use the data from the previous example in this section, and test the hypothesis that the proportions are equal against the alternative that $\pi_1 > \pi_2$.

(1) We are to test $H_o: \pi_1 - \pi_2 = 0$ against $H_1: \pi_1 - \pi_2 > 0$.

(2) Let $\alpha = 0 \cdot 05$; $z_{0.05} = 1 \cdot 645$.

(3) The samples have been obtained. We must assume that they are random samples from binomial populations.

(4)
$$x_1 = 1078, \ n_1 = 3241, \ p_1 = \frac{1078}{3214} = 0 \cdot 335$$
$$x_2 = 543, \ n_2 = 2011, \ \hat{p}_2 = 0 \cdot 270$$
$$\hat{p} = \frac{1078 + 543}{3214 + 2011} = \frac{1621}{5225} = 0 \cdot 310$$

(5)
$$z = \frac{(0 \cdot 335 - 0 \cdot 270) - 0}{\sqrt{\dfrac{(0 \cdot 310)(0 \cdot 690)}{3214} + \dfrac{(0 \cdot 310)(0 \cdot 690)}{2011}}}$$
$$= \frac{0 \cdot 065}{\sqrt{0 \cdot 0000666 + 0 \cdot 0001064}} = \frac{0 \cdot 065}{\sqrt{0 \cdot 0001730}}$$
$$= \frac{0 \cdot 065}{0 \cdot 013} = 5 \cdot 00$$

(6) $5 \cdot 00 > 1 \cdot 645$. Reject H_o.

TESTS ABOUT THE DIFFERENCE OF TWO MEANS

Case 1. Normal Populations, Unknown Common Variance

The t-distribution can be used to test a hypothesis about the difference between the means of two normal populations if the variances of the populations are equal.

More specifically, if $\bar{x}_1$ is the mean of a sample of size n_1 from a normal population with mean μ_1 and variance σ^2, and if $\bar{x}_2$ is the mean of a sample of size n_2 from a normal population with mean μ_2 and variance σ^2, the random variable

$$t = \frac{(\bar{x}_1 - \bar{x}_2) - (\mu_1 - \mu_2)}{\sqrt{s_p^2 \left(\dfrac{1}{n_1} + \dfrac{1}{n_2} \right)}} \tag{8.10}$$

where

$$s_p^2 = \frac{(n_1 - 1)s_1^2 + (n_2 - 1)s_2^2}{n_1 + n_2 - 2} \qquad (8.11)$$

has a t-distribution with $n_1 + n_2 - 2$ degrees of freedom.

The quantity s_p^2 is known as the **pooled sample variance**. Because the variances of the two populations are assumed to be equal, each of the separate sample variances is an estimate of the same population variance. Then, under this assumption, we should combine (pool) the two sample variances and get a single estimate of the common population variance. This pooled estimate is given in Formula (8.11); it is a weighted average of the two separate sample variances.

The numerical value of the random variable s_p^2 that is determined by two-particular random samples is

$$s_p^2 = \frac{(n_1 - 1)s_1^2 + (n_2 - 1)s_2^2}{n_1 + n_2 - 2} \qquad (8.12)$$

where s_1^2 is the variance of the first sample and s_2^2 is the variance of the second sample. Formula (8.12) is equivalent to

$$s_p^2 = \frac{(\text{Sum of squares from Sample 1}) + (\text{Sum of squares from Sample 2})}{(\text{Degrees of freedom from Sample 1}) + (\text{Degrees of freedom from Sample 2})} \qquad (8.13)$$

In words, to obtain the pooled sample variance we 'pool' (that is, we add) the sums of squares and we 'pool' the degrees of freedom; then we divide the pooled sum of squares by the pooled degrees of freedom. The usual computing formulae are used to find the sums of squares.

The assumptions that must be made in order to use the t-distribution to test a hypothesis about the difference between two population means are:

(1) the two populations are normal;
(2) the two populations have the same variance; and
(3) the two samples are random ones.

If an experimenter wants to test assumption (2) he can use the F-distribution.

The hypothesis that is usually tested is that the means of two normal populations are equal, $H_0: \mu_1 = \mu_2$. This is the same as testing that the difference between the two means is zero, $H_0: \mu_1 - \mu_2 = 0$. However, we might also want to test that μ_1 is three units larger than μ_2, $H_0: \mu_1 - \mu_2 = 3$; or that μ_1 is 7 units larger than μ_2, $H_0: \mu_1 - \mu_2 = 7$. In general, then, we might wish to test that the difference between μ_1 and μ_2 is some specified number, say Δ_0. Any hypothesis about the mean of two populations that we might want to test can be written in the form $H_0: \mu_1 - \mu_2 = \Delta_0$. The experimenter can denote as Population 1 whichever population he wishes.

Procedure for Testing $H_0: \mu_1 - \mu_2 = \Delta_0$

(1) Formulate the null and alternative hypotheses:

$$H_0: \mu_1 - \mu_2 = \Delta_0 \quad \text{versus} \quad H_1: \mu_1 - \mu_2 \neq \Delta_0$$

Statistics Made Simple

(2) Decide on an α-level.

Look up $t_{\alpha/2}(n_1 + n_2 - 2)$.

The critical region consists of all *t*-values greater than or equal to $t_{\alpha/2}(n_1 + n_2 - 2)$ and of all *t*-values less than or equal to $-t_{\alpha/2}(n_1 + n_2 - 2)$. See Fig. 8.4.

(3) Obtain the two random samples.

Calculate $\bar{x}_1$ and $\bar{x}_2$.

Calculate s_p^2. The easiest way to do this is to first calculate the two sums of squares for the two samples:

$$\sum (x_{1i} - \bar{x}_1)^2 = \sum x_{1i}^2 - \frac{(\sum x_{1i})^2}{n_1}$$

$$\sum (x_{2i} - \bar{x}_2)^2 = \sum x_{2i}^2 - \frac{(\sum x_{2i})^2}{n_2}$$

Then calculate $\quad s_p^2 = \dfrac{\sum (x_{1i} - \bar{x}_1)^2 + \sum (x_{2i} - \bar{x}_2)^2}{n_1 + n_2 - 2}$

(4) Calculate $\quad t = \dfrac{(\bar{x}_1 - \bar{x}_2) - \Delta_0}{\sqrt{s_p^2 \left(\dfrac{1}{n_1} + \dfrac{1}{n_2} \right)}}$

(5) Note whether *t* is in the critical region and accept or reject H_o.

EXAMPLE: The following data are the gains in mass, measured in kilogrammes, of babies from birth to age one year. All babies in both groups weighed approximately the same at birth. The babies in Sample 1 were fed formula A, and the babies in Sample 2 were fed formula B. (Assume that the experimenter has no preconceived notions about which formula might be better.)

Sample 1	Sample 2
5	9
7	10
8	8
9	6
6	8
7	7
10	9
8	
6	

Test at the 5 per cent level of significance that the mean of Population 1 equals the mean of Population 2.

(1) H_0: $\mu_1 - \mu_2 = 0$ versus H_1: $\mu_1 - \mu_2 \neq 0$. As previously stated, testing that the difference between the population means equals zero is equivalent to testing that the population means are equal.

A two-tailed alternative rather than a one-tailed alternative is selected because, prior to performing the experiment, we do not suspect which formula causes more weight to be gained (if either does).

(2) The α-value was given:

$$\alpha = 0.05$$

$$t_{0.025}(9 + 7 - 2) = t_{0.025}(14) = 2.145$$

The critical region consists of all values of *t* less than or equal to -2.145 or greater than or equal to 2.145.

(3) The random samples are given.
The sum of squares for the first sample is

$$\sum (x_{1i} - \bar{x}_1)^2 = \sum x_{1i}^2 - \frac{\left(\sum x_{1i}\right)^2}{n_1}$$

$$= 504 - \frac{(66)^2}{9}$$

$$= 504 - \frac{4,356}{9}$$

$$= 504 - 484$$

$$= 20$$

The sum of squares for the second sample is

$$\sum (x_{2i} - \bar{x}_2)^2 = \sum x_{2i}^2 - \frac{\left(\sum x_{2i}\right)^2}{n_2}$$

$$= 475 - \frac{(57)^2}{7}$$

$$= 475 - \frac{3\,249}{7}$$

$$= 475 - 464 \cdot 1$$

$$= 10 \cdot 9$$

The pooled sample variance is given by

$$s_p^2 = \frac{20 + 10 \cdot 9}{14}$$

$$= 2 \cdot 21$$

The sample means are

$$\bar{x}_1 = \frac{66}{9} = 7 \cdot 33$$

and

$$\bar{x}_2 = \frac{57}{7} = 8 \cdot 14$$

(4)

$$t = \frac{(7 \cdot 33 - 8 \cdot 14) - 0}{\sqrt{2 \cdot 21 \left(\frac{1}{9} + \frac{1}{7}\right)}}$$

$$= \frac{-0 \cdot 81}{\sqrt{(2 \cdot 21)(0 \cdot 254)}}$$

$$= \frac{-0 \cdot 81}{\sqrt{0 \cdot 56134}}$$

$$\approx \frac{-0 \cdot 81}{0 \cdot 7492}$$

$$= -1 \cdot 0812$$

(5) $-1 \cdot 0812 > -2 \cdot 145$

Accept H_o, since $-1 \cdot 0812$ is not in the critical region.

If a one-sided alternative hypothesis is desired the only changes which must be made in the above procedure are in steps (1) and (2).

When the alternative hypothesis is H_1: $\mu_1 - \mu_2 > \Delta_0$ the critical region is the right-hand tail of the t-distribution (see Fig. 8.2). Then step (2) should be changed to read as follows:

(2) Decide on an α-level.
Look up $t_\alpha(n_1 + n_2 - 2)$.
The critical region consists of all t-values greater than or equal to $t_\alpha(n_1 + n_2 - 2)$.

When the alternative hypothesis is $H_1: \mu_1 - \mu_2 < \Delta_0$, the critical region is the left-hand tail of the t-distribution (see Fig. 8.3). Then step (2) should be changed to read as follows:

(2) Decide on an α-level.
Look up $t_\alpha(n_1 + n_2 - 2)$.
The critical region consists of all t-values less than or equal to

$$-t_\alpha(n_1 + n_2 - 2)$$

EXAMPLE: Given the following samples, test $H_o: \mu_1 - \mu_2 = 3$ versus $H_1: \mu_1 - \mu_2 > 3$. Let $\alpha = 0.10$.

Sample 1: 51, 42, 49, 55, 46, 63, 56, 58, 47, 39, 47
Sample 2: 38, 49, 45, 29, 31, 35

(1) We are testing $H_o: \mu_1 - \mu_2 = 3$ versus $H_1: \mu_1 - \mu_2 > 3$.
(2) We are given $\alpha = 0.10$; $n_1 + n_2 - 2 = 6 + 11 - 2 = 15$; $t_{0.10}(15) = 1.341$.
The critical region consists of all values of t greater than or equal to 1.341.
(3) The random samples are given.

The sample means are $\qquad \bar{x}_1 = \dfrac{553}{11} = 50.3$

and $\qquad \bar{x}_2 = \dfrac{227}{6} = 37.8$

The two sums of squares are:

$$\sum (x_{1i} - \bar{x}_1)^2 = 28\ 315 - \frac{(553)^2}{11}$$

$$= 28\ 315 - \frac{305\ 809}{11}$$

$$= 28\ 315 - 27\ 801$$

$$= 514$$

$$\sum (x_{2i} - \bar{x}_2)^2 = 8897 - \frac{(227)^2}{6}$$

$$= 8897 - \frac{51\ 529}{6}$$

$$= 8897 - 8588$$

$$= 309$$

The pooled variance is $\qquad s_p^2 = \dfrac{514 + 309}{15}$

$$= \frac{823}{15}$$

$$= 54.9$$

(4) $\qquad\qquad t = \dfrac{(50.3 - 37.8) - 3}{\sqrt{54.9(\frac{1}{6} + \frac{1}{11})}}$

$$= \frac{9.5}{\sqrt{14.14}}$$

$$= \frac{9.5}{3.76}$$

$$= 2.53$$

(5) $2.55 > 1.341$
The value of t is in the critical region. Reject H_o and accept H_1.

Case 2. Normal Populations, Known Variances

When the variances of two normal populations are known, the random variable

$$z = \frac{(\bar{x}_1 - \bar{x}_2) - (\mu_1 - \mu_2)}{\sqrt{\sigma_1^2/n_1 + \sigma_2^2/n_2}}$$

is exactly standard normal, where $\bar{x}_1$ is the mean of a sample of size n_1 from Population 1 and $\bar{x}_2$ is the mean of a sample of size n_2 from Population 2. For two particular samples, the quantity

$$z = \frac{(\bar{x}_1 - \bar{x}_2) - \Delta_0}{\sqrt{\sigma_1^2/n_1 + \sigma_2^2/n_2}}$$

is calculated, referred to normal tables, and the null hypothesis $H_0: \mu_1 - \mu_2 = \Delta_0$ is accepted or rejected. The procedure is stated more formally below.

Procedure for Testing $H_0: \mu_1 - \mu_2 = \Delta_0$ versus a Specified Alternative

(1) Formulate the null and alternative hypotheses:
$H_0: \mu_1 - \mu_2 = \Delta_0$ versus the specified alternative.
(2) Decide upon the α-level, look up the appropriate values from the standard normal table, and note the critical region.
(3) Obtain a random sample of size n_1 from Population 1 and a random sample of size n_2 from Population 2.
Calculate $\bar{x}_1$ and $\bar{x}_2$.
(4) Calculate $z = \dfrac{(\bar{x}_1 - \bar{x}_2) - \Delta_0}{\sqrt{\sigma_1^2/n_1 + \sigma_2^2/n_2}}$
(5) Note whether z is in the critical region, and accept or reject H_0.

Case 3. Non-normal Populations, Known Variances

When the sample sizes are large (more than 30, say) it does not matter whether the populations are normal or not. Regardless of the form of the distribution, if the sample sizes are large the random variable

$$z = \frac{(\bar{x}_1 - \bar{x}_2) - (\mu_1 - \mu_2)}{\sqrt{\sigma_1^2/n_1 + \sigma_2^2/n_2}}$$

is approximately standard normal.

Procedure for Testing $H_0: \mu_1 - \mu_2 = \Delta_0$ versus a Specified Alternative

Exactly the same as the procedure under Case 2.

Case 4. Non-normal Populations, Unknown Variances

When the sample sizes are large, the random variable

$$z = \frac{(\bar{x}_1 - \bar{x}_2) - (\mu_1 - \mu_2)}{\sqrt{s_1^2/n_1 + s_2^2/n_2}}$$

is approximately normal, no matter what the form of the distributions or whether their variances are known.

Procedure for Testing H_0: $\mu_1 - \mu_2 = \Delta_0$ *versus a Specified Alternative*

The same as the procedure under Case 2, excepting steps (3) and (4), which are:

(3) Obtain a random sample of size n_1 from Population 1 and a random sample of size n_2 from Population 2.
Calculate $\bar{x}_1$ and $\bar{x}_2$.
Calculate s_1^2 and s_2^2.

(4) Calculate $z = \dfrac{(\bar{x}_1 - \bar{x}_2) - \Delta_0}{\sqrt{s_1^2/n_1 + s_2^2/n_2}}$

EXAMPLE: A sample of size 30 from a non-normal population yielded the sample values $\bar{x}_1 = 80$, $s_1^2 = 150$. A sample of size 40 from a second non-normal population yielded the sample values $\bar{x}_2 = 71$, $s_2^2 = 200$. Test H_0: $\mu_1 - \mu_2 = 2$, versus H_1: $\mu_1 - \mu_2 > 2$.

(1) H_0: $\mu_1 - \mu_2 = 2$, against H_1: $\mu_1 - \mu_2 > 2$.
(2) Let $\alpha = 0.05$; $z_{0.05} = 1.645$; the critical region is $z > 1.645$.
(3) The sample values are given to be

$$\bar{x}_1 = 80, \ \bar{x}_2 = 71$$
$$s_1^2 = 150, \ s_2^2 = 200$$

(4)
$$z = \frac{80 - 71 - 2}{\sqrt{\dfrac{150}{30} + \dfrac{200}{40}}} = \frac{7}{\sqrt{5 + 5}} = 2.21$$

(5) We see that $z = 2.21$ is in the critical region. Therefore, reject H_0.

EXERCISES

1. A sample of size 25 from a normal population yielded the sample values $\bar{x} = 18.1$ and $s^2 = 16$. Test H_0: $\mu = 16$ versus H_1: $\mu > 16$. Let $\alpha = 0.01$.

2. Given that the observations 9, 11, 11, 14 have been drawn from a normal population. Test H_0: $\mu = 8.8$ against H_1: $\mu \neq 8.8$. Let $\alpha = 0.05$.

3. Given that a sample of size 100 yielded $\bar{x} = 234$ and $s^2 = 400$. Test H_0: $\mu = 250$ versus H_1: $\mu \neq 250$. Let $\alpha = 0.10$.

4. Test H_0: $\mu = 150$ versus H_1: $\mu > 150$ if a sample of size 36 yielded $\bar{x} = 159$. It is known that $\sigma^2 = 400$.

5. A candidate for mayor in a large city believes that he appeals to at least 10 per cent more of the women voters than the men voters. He hires the services of a poll-taking organization, and they find that 62 of 100 women interviewed support the candidate, and 69 of 150 men support him. At the 0.05 level is the hypothesis accepted or rejected?

6. A book club selected a random sample from those persons on its membership list that had purchased a certain recent club selection. Each member of the sample was asked whether they were male or female, and if they had read the selection, whether they had liked it. Of the 1500 respondents, 1000 were women and 500 were men. Of the women, 850 had liked the book; and 400 of the men had liked it. Is there a difference between the proportion of women and the proportion of men who liked the book?

7. A sample of size 6 from a normal population with variance 24 gave $\bar{x}_1 = 15$. A sample of size 8 from a normal population with variance 80 gave $\bar{x}_2 = 13$. Test H_0: $\mu_1 - \mu_2 = 0$ against H_1: $\mu_1 - \mu_2 \neq 0$. Let $\alpha = 0.05$.

8. A sample of size 6 from a normal population gave $\bar{x}_1 = 30$ and $s_1^2 = 40$. A sample of size 11 from a normal population gave $\bar{x}_2 = 22$ and $s_2^2 = 45$. Test H_0: $\mu_1 - \mu_2 = 4$ versus H_1: $\mu_1 - \mu_2 > 4$. Let $\alpha = 0.05$.

CHAPTER IX

CORRELATION AND REGRESSION

THE SAMPLE CORRELATION COEFFICIENT

When we select a random sample of size n and make two observations on each member of the sample we then have n *pairs* of observations. For instance, we might select a random sample of people from a population and measure height and weight, or I.Q. and earnings, or age and the score on a test that measures adjustment to authority. Or, for a sample of rats, we might measure (for each rat) the time since the last feeding and the time it takes to traverse a maze with food at the end. Or, for a sample of plots of wheat, we might measure the yield of each plot and the millimetres of rainfall which fell on each plot (assuming similar soil and climate conditions). All of these situations have something in common—each member of the random sample has two attributes of interest that can be measured.

We will denote the measurement of the first attribute by x, and the measurement of the second by y. The pair of observations on the first individual would be (x_1, y_1); the observations on the second one, (x_2, y_2); and so on. Then the random sample consists of n pairs of observations, $(x_1, y_1), (x_2, y_2), \ldots (x_n, y_n)$.

Each pair of x and y values can be considered as the x and y coordinates of a point, and these points can be plotted in the usual way on a pair of axes. Suppose, for example, that a random sample is selected from the population of men between the ages of 40 and 50 in a certain city who are employed full-time, and the number of years of schooling (x) and the annual income in thousands of £ (y) recorded for each man. Suppose further that the random sample of 12 men yielded the following data:

Table 9.1

Schooling (years) (x)	Income (thousands of £) (y)
10	6
7	4
12	7
12	8
9	10
16	7
12	10
18	15
8	5
12	6
14	11
16	13

In this chapter we are exclusively interested in determining how x and y are related *linearly*. There are various other sorts of relationships, but we will not consider them.

A diagram like that in Fig. 9.1 is called a **scatter diagram**. It gives a rough idea of how the variables x and y are related. We cannot draw defensible conclusions by merely examining data, whether we examine the sample observations themselves or a representation of the sample observations, such as a histogram or a scatter diagram. In other words, we cannot simply look at Fig. 9.1 and conclude that since more than half of the points appear to be nearly in a straight line there is a linear relationship between x and y. On the other hand, neither can we conclude that there is *not* a linear relationship.

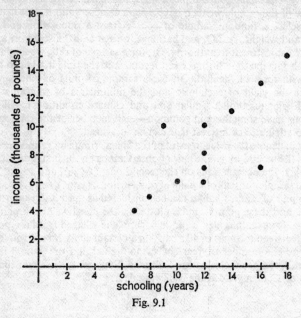

Fig. 9.1

We need a quantity that yields a *number*—a number that is a measure of how much linear relationship there is between x and y. The quantity that is used for this purpose is known as the **sample correlation coefficient**, denoted by r, and is given by the formula

$$r = \frac{\sum(x - \bar{x})(y - \bar{y})}{\sqrt{\left[\sum(x - \bar{x})^2\right]\left[\sum(y - \bar{y})^2\right]}} \tag{9.1}$$

This quantity can be written in various other forms that are algebraically equivalent or, in some cases, symbolically equivalent.

The value of the sample correlation coefficient is an estimate of a theoretical quantity known as the **population correlation coefficient**, denoted by ρ. (Rho, written ρ, and pronounced 'roe', is the seventeenth letter of the Greek alphabet.) The value of ρ is always between -1 and $+1$.

The value of the sample correlation coefficient, also, is always between

−1 and +1. A value of *r* equal to −1 indicates a perfect linear relationship between the sample values of *x* and *y*, with the value of *y* *decreasing* as the value of *x* *increases*—the larger *x* becomes, the smaller *y* becomes; and the smaller *x* becomes, the larger *y* becomes. A value of *r* equal to +1 also indicates a perfect linear relationship between the sample values, but one in which the value of *y* increases as *x* increases. Larger values of *y* are associated with larger values of *x*; and smaller values of *y* are associated with smaller values of *x*. If there is no linear relationship between the sample values of *x* and *y*, then *r* will have a value near zero. As *r* increases from 0 to +1 (or decreases from 0 to −1) the linear relationship between the sample values of *x* and *y* becomes more pronounced.

The scatter diagrams of several samples, along with the values of *r* that the sample yielded, are shown in Fig. 9.2.

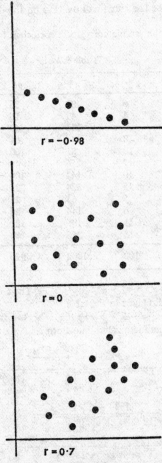

Fig. 9.2

COMPUTATION OF r

A convenient computing form of Formula (9.1) is

$$r = \frac{\sum xy - \frac{(\sum x)(\sum y)}{n}}{\sqrt{\left[\sum x^2 - \frac{(\sum x)^2}{n}\right] \cdot \left[\sum y^2 - \frac{(\sum y)^2}{n}\right]}} \qquad (9.2)$$

Examining Formula (9.2) we see that in order to compute r it is necessary to find five sums $-\sum x$, $\sum y$, $\sum x^2$, $\sum y^2$, and $\sum xy$. On standard rotary desk calculators it is usually possible to obtain all five of these sums in one sequence of operations. Occasionally, however, if the sample is large, or if the values of x or y or both are large, it will not be possible to obtain all of these sums simultaneously because the sums will overlap on the dials of the machine.

EXAMPLE: Compute the sample correlation coefficient for the data in Table 9.2 using Formula 9.2.

Table 9.2

x	y	x^2	y^2	xy
10	6	100	36	60
7	4	49	16	28
12	7	144	49	84
12	8	144	64	96
9	10	81	100	90
16	7	256	49	112
12	10	144	100	120
18	15	324	225	270
8	5	64	25	40
12	6	144	36	72
14	11	196	121	154
16	13	256	169	208
$\sum$ 146	102	1902	990	1334

We need the sums $\sum x$, $\sum y$, $\sum x^2$, $\sum y^2$, $\sum xy$. The computations necessary to obtain these sums are displayed in the table.

We now have $\sum x = 146$, $\sum y = 102$, $\sum x^2 = 1902$, $\sum y^2 = 990$, and $\sum xy = 1334$. Substituting these values into Formula (9.2), we have

$$r = \frac{1334 - \frac{(146)(102)}{12}}{\sqrt{\left[1902 - \frac{(146)^2}{12}\right] \cdot \left[990 - \frac{(102)^2}{12}\right]}}$$

$$= \frac{1334 - 1241}{\sqrt{[1902 - 1776] \cdot [990 - 867]}}$$

$$= \frac{93}{\sqrt{(126)(123)}} = \frac{93}{\sqrt{15\,498}} = \frac{93}{124 \cdot 5} = 0 \cdot 747$$

TESTING HYPOTHESES ABOUT THE POPULATION CORRELATION COEFFICIENT

Recall that we have previously stated that r is an estimate of ρ. Therefore, we can use r to test hypotheses about ρ.

The random variable $\mathbf{r}$, particular numerical values of which are given by Formula (9.1), has neither a normal distribution nor a distribution that becomes approximately normal as the sample size increases. However, $\mathbf{r}$ can be transformed into another random variable that *is* approximately normal. If $\mathbf{x}$ and $\mathbf{y}$, particular numerical values of which appear in Formula (9.1), are normal random variables (actually, if $\mathbf{x}$ and $\mathbf{y}$ together have what is known as a **bivariate normal distribution**), it can be shown by advanced methods that the random variable $\mathbf{z}_f$, where

$$\mathbf{z}_f = (1 \cdot 1513) \cdot \log \frac{1 + \mathbf{r}}{1 - \mathbf{r}} \tag{9.3}$$

is approximately normal with mean

$$\mu_z = (1 \cdot 1513) \cdot \log \frac{1 + \rho}{1 - \rho} \tag{9.4}$$

and variance $$\sigma_z^2 = \frac{1}{n - 3} \tag{9.5}$$

The random variable $(1 \cdot 1513) \log \dfrac{1 + \mathbf{r}}{1 - \mathbf{r}}$ is denoted $\mathbf{z}_f$ because this transformation that changes $\mathbf{r}$ into $\mathbf{z}_f$ is called the Fisher-z transformation, in honour of the statistician who discovered it

$$\left[\text{The symbols } \log \frac{1 + \rho}{1 - \rho} \text{ denote the common logarithm (base 10) of } \frac{1 + \rho}{1 - \rho} \right]$$

(See *Mathematics Made Simple*, Chapter 12.)

In a particular sample the random variable $\mathbf{r}$ has the numerical value r given by Formula (9.1); the corresponding numerical value of the random variable $\mathbf{z}_f$ is z_f, where

$$z_f = (1 \cdot 1513) \cdot \log \frac{1 + r}{1 - r} \tag{9.6}$$

We will use z_f, rather than r, in order to test a hypothesis about ρ. Fortunately, it is not necessary to use Formula (9.6) in order to find the z_f that corresponds to a particular r. There is a table in the Appendix that gives the z_f values that correspond to r values between 0·00 and 0·99. (Before using the table, r should be rounded to two decimal places.) The table is entered by means of the two digits of r—the first digit is found along the left-hand margin of the table and the second digit at the top.

EXAMPLE: Find the z_f that corresponds to $r = 0 \cdot 73$. The first digit of r is 7. Read down the column of first digits at the left-hand margin until 0·7 is reached. Then read across that row until the column head by 0·03 is reached. At this point we read 0·92873 as the z_f that corresponds to $r = 0 \cdot 73$.

EXAMPLE: Find the z_f that corresponds to $r = 0 \cdot 48$. Read down the column of first digits until 0·4 is reached. Then read across that row until the column headed by 0·08 is reached. We read $z_f = 0 \cdot 52298$.

Some other z_f-values are found below. The reader should verify these for practice:

$$r = 0.07; \qquad z_f = 0.07012$$
$$r = 0.55; \qquad z_f = 0.61838$$
$$r = 0.80; \qquad z_f = 1.09861$$

The quantity r can be negative as well as positive. The z_f that corresponds to a negative r is equal to the negative of the z_f that corresponds to the absolute value of that r (for a discussion of absolute value see Chapter IV). For example, to find the z_f that corresponds to $r = -0.41$, find the z_f that corresponds to $r = 0.41$, and make it negative; to find the z_f that corresponds to $r = -0.15$, find the z_f that corresponds to $r = 0.15$, and make it negative.

EXAMPLE: Find the z_f that corresponds to $r = -0.26$. We find that the z_f that corresponds to $r = 0.26$ is 0.26611. Therefore, the z_f that corresponds to $r = -0.26$ is -0.26611.

The reader should verify the following values of z_f if he feels that he needs more practice:

$$\text{If } r = -0.19, \text{ then } z_f = -0.19234$$
$$\text{If } r = -0.83, \text{ then } z_f = -1.18813$$
$$\text{If } r = -0.35, \text{ then } z_f = -0.36544$$

The r that corresponds to a given z_f can also be easily found. (The ability to find the r that corresponds to a given z_f is not needed until the next chapter.) Usually a z_f will be given (or calculated) to only three decimal places, although the z_f-values in the table are correct to five places.

EXAMPLE: Find the r that corresponds to $z_f = 0.892$.
We search in the tables for the z_f that is nearest to 0.89200. The nearest z_f is 0.88718. The r that corresponds to 0.88718 is 0.71. The r correct to the nearest hundredth that corresponds to $z_f = 0.892$ is 0.71.

EXAMPLE: Find the r that corresponds to $z_f = -1.345$.
The value nearest to $z_f = 1.345$ is $z_f = 1.33308$. The r correct to the nearest hundredth that corresponds to $z_f = 1.345$ is 0.87.
The r that corresponds to $z_f = -1.345$ is -0.87.

Since the formula for the mean of z_f, $\mu_z = (1.1513) \cdot \log \dfrac{1 + \rho}{1 - \rho}$, is analogous to the formula for z_f, $z_f = (1.1513) \cdot \log \dfrac{1 + r}{1 - r}$, the table can also be used to find the mean of z_f that corresponds to a value of the theoretical ρ. For example, if $\rho = 0.43$, then $\mu_z = 0.45990$.

As previously stated, the random variable z_f is approximately normal with mean μ_z and variance σ_z^2, where μ_z and σ_z^2 are given by Formula (9.4) and (9.5). Recall that a variable is standardized by subtracting its mean from itself and dividing the resulting difference by its standard deviation. If an approximately normal random variable is standardized the resulting variable has a distribution that is approximately standard normal. Therefore, the standardized random variable $\dfrac{z_f - \mu_z}{\sigma_z}$ is approximately standard normal. Thus we have a method to test hypotheses about ρ: we compare the numerical quantity

$\dfrac{z_f - \mu_z}{\sigma_z}$ with the appropriate critical value (or values) from standard normal tables. The procedure is outlined and illustrated below.

Procedure for Testing H_0: $\rho = \rho_0$ versus H_1: $\rho \neq \rho_0$

NOTE: Do not confuse the various symbols in this procedure: z_f denotes the Fisher-z value, given by Formula (9.6) and found from the Fisher-z table; $z_{\alpha/2}$ denotes the upper $\alpha/2$-point of the standard normal distribution (z_α is defined similarly); z denotes the statistic that is found in step (4) below, specifically, $z = (z_f - \mu_z)\sqrt{n-3}$.

(1) Formulate the null and alternative hypotheses.
(2) Decide on the value for α, and find $z_{\alpha/2}$ in standard normal tables.
    The critical region is $z \leqslant -z_{\alpha/2}$ and $z \geqslant z_{\alpha/2}$.
(3) Obtain a random sample and measure x and y. (In nearly all textbook problems the values of x and y are given.)
    Calculate r, using Formula (9.2).
(4) Find, in the Fisher-z table, the z_f that corresponds to r.
    Find, in the Fisher-z table, the μ_z that corresponds to ρ_0.

Calculate $z = \dfrac{z_f - \mu_z}{\sqrt{\dfrac{1}{n-3}}} = (z_f - \mu_z)\sqrt{n-3}$

(5) Reject H_0 if z is in the critical region.

The procedure for testing H_0: $\rho = \rho_0$ against the one-sided alternative H_1: $\rho < \rho_0$ is the same as that above except for step (2), which should be modified as follows:

(2) Decide on a value for α, and find z_α in standard normal tables.
    The critical region is $z \leqslant -z_\alpha$.

The procedure for testing H_0: $\rho = \rho_0$ against the one-sided alternative H_1: $\rho > \rho_0$ is the same as that above except for step (2), which should be modified as follows:

(2) Decide on a value for α, and find z_α in standard normal tables.
    The critical region is $z \geqslant z_\alpha$.

EXAMPLE: Given the data in Table 9.1, test the hypothesis that $\rho = 0.6$ versus the one-sided alternative that $\rho > 0.6$.

(1) We are testing H_0: $\rho = 0.6$ versus H_1: $\rho > 0.6$.
(2) We will choose $\alpha = 0.05$; $z_{0.05} = 1.65$. The critical region is $z \geqslant 1.65$.
(3) The value of r was computed in the previous section: $r = 0.75$, correct to two places.
(4) The z_f that corresponds to $r = 0.75$ is $z_f = 0.97295$. The μ_z that corresponds to $\rho = 0.6$ is $\mu_z = 0.69315$. Rounding the z_f and μ_z values to three places, we have

$$(z_f - \mu_z)\sqrt{n-3} = \sqrt{9}(0.973 - 0.693)$$
$$= 3(0.280)$$
$$= 0.840$$

(5) 0.840 is not in the critical region. Accept H_0.

EXAMPLE: The data in Table 9.3 represent the scores (perfect score, 75) on a mathematics placement test and the final averages (perfect score, 100) in a mathematics course for a random sample of eleven students. Test whether the placement test can be used to help predict which students will perform well in mathematics.

Table 9.3

Student	Placement test score (x)	Final average in mathematics (y)
1	51	75
2	52	72
3	59	82
4	45	67
5	61	75
6	54	79
7	56	78
8	67	82
9	63	87
10	53	72
11	60	96

(1) The null hypothesis will be that there is no correlation between placement test scores and mathematics grades: $H_0: \rho = 0$. It seems reasonable to expect students who did poorly on the placement test to do poorly in mathematics, and students who did well on the placement test to do well in mathematics. (The terms 'poorly' and 'well' are being used as relative terms. If a student does poorly, relative to the scores of others, on the placement test, we would reasonably expect him to do poorly, relative to the averages of others, in mathematics.) Therefore, assuming that the relationship that exists between placement test scores and final averages is linear, we would expect the population correlation coefficient to be positive. That is, the natural alternative hypothesis is that there is a positive linear relationship between the x and y: $H_1: \rho > 0$. We will test $H_0: \rho = 0$ versus $H_1: \rho > 0$.

(2) We will choose 0·01 for the level of significance of this test; $z_{0.01} = 2.33$. The critical region is $z \geqslant 2·33$.

(3) We must first calculate r. The pencil-and-paper calculations necessary to find the various sums in the computing formula for r are shown in Table 9.4.

Table 9.4

x	y	x^2	y^2	xy
51	75	2 601	5 625	3 825
52	72	2 704	5 184	3 744
59	82	3 481	6 724	4 838
45	67	2 025	4 489	3 015
61	75	3 721	5 625	4 575
54	79	2 916	6 241	4 266
56	78	3 136	6 084	4 368
67	82	4 489	6.724	5 494
63	87	3 969	7 569	5 481
53	72	2 809	5 184	3 816
60	96	3 600	9 216	5 760
$\sum$ 621	865	35 451	68 665	49 182

We thus have $\sum x = 621$, $\sum y = 865$, $\sum x^2 = 35\,451$, $\sum y^2 = 68\,665$, and $\sum xy = 49\,182$. Substituting these sums into Formula 9.2, we have

$$r = \frac{49\,182 - \dfrac{(621)(865)}{11}}{\sqrt{\left[35\,451 - \dfrac{(621)^2}{11}\right] \cdot \left[68\,665 - \dfrac{(865)^2}{11}\right]}}$$

which yields
$$r = 0 \cdot 69$$

The correlation coefficient is unchanged if any number is subtracted from (or added to) each x; if any number is subtracted from (or added to) each y; or if a number is subtracted from (or added to) each x and a (usually different) number is subtracted from (or added to) each y. (Similar statements can be made about multiplication.) For example, r is unchanged if we subtract 10 from each x and multiply each y by 5; or if we multiply each x by 2 and leave y unchanged; or if we add 500 to each x and subtract 75 from each y; and so on. The x and y values can frequently be changed into other values that are more convenient for computational purposes.

(4) The z_f that corresponds to $r = 0 \cdot 69$ is $z_f = 0 \cdot 848$ (correct to three places).
    The μ_z value that corresponds to $\rho = 0$ is $\mu_z = 0$.
    The value of the test statistic is $\sqrt{8}(0 \cdot 848) = 2 \cdot 40$.
(5) $2 \cdot 40 > 2 \cdot 33$.
    Reject H_0: $\rho = 0$ and accept H_1: $\rho > 0$.

LINEAR REGRESSION

The sample correlation coefficient is a measure of the linear relationship between two variables x and y, but it does not tell us what the relationship is. It is often desirable to determine the relationship in order to predict the value of one variable from that of the other. Finding the equation that gives the relationship between y and x is known as **finding the regression line.** The determination of the regression line is discussed in the next section.

The point of view is somewhat different in a regression problem from that in a correlation problem. In the regression approach the variable x is not a random variable; it is assumed that its values can be chosen by the experimenter and measured to whatever degree of accuracy he desires (subject to the limitations of the device that is being used to measure x, of course). This is different from the correlation approach, in which both variables must be random variables. However, the regression approach can be used on data that have been selected in anticipation of using the correlation approach.

For instance, if we want to find the coefficient of correlation between heights (x) and weights (y) for a particular body type we select a random sample of such individuals. The height and weight of each individual is measured. Both height and weight are random variables. It is impossible to predict in advance how the sample of heights will be distributed, and it is impossible to predict in advance how the sample of weights will be distributed. The values of both x and y are random quantities, and different samples will yield different sets of x-values and y-values.

On the other hand, if we want to find the regression line that gives the linear relationship between heights and weights it is permissible to select certain heights. The experimenter might want to make certain that he has the same number of observations for each height—for instance, three observations for a height of 1·90 metres, three for a height of 1·91 metres, and so on. Thus, since the *x*-values may be known in advance, they are not random variables. The *y*-values continue to be random variables, however—it cannot be predicted in advance which particular set of *y*-values will be the result of a particular experiment, nor which *y*-value(s) will correspond to a particular *x*-value.

FINDING THE REGRESSION (LEAST-SQUARES) LINE

Two topics are treated in this section. First, we will discuss how to find the equation of a straight line. Usually this topic is first studied in school algebra, or perhaps the reader is already familiar with it from *Additional Mathematics Made Simple*, Chapter 6; if so this material may be omitted with no loss of continuity. Second, we will learn how to find the equation of the regression line which best fits the sample data.

The equation of a straight line can be given in terms of its *y*-intercept and its slope. The **y-intercept** of a straight line is the *y*-co-ordinate of the point where the line crosses the *y*-axis.

We will denote the line whose equation is $y = 1 + x$ by L_1; the line whose equation is $y = 1 - \frac{1}{2}x$ by L_2; and the line whose equation is $y = -2 + 4x$ by L_3.

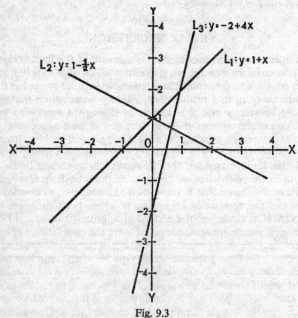

Fig. 9.3

In Fig. 9.3 the *y*-intercepts of L_1 and L_2 are both 1; the *y*-intercept of L_3 is -2.

The **slope of a line** is the amount of change in *y* per unit change in *x*: a slope of 2 means that *y* increases 2 units while *x* increases 1 unit; a slope of $\frac{1}{3}$ means that *y* increases $\frac{1}{3}$ unit while *x* increases 1 unit (equivalently, *y* increases

1 unit while x increases 3 units); and a slope of -1 means that y *decreases* 1 unit while x increases 1 unit. An equivalent definition of the slope of a line is to define it as the ratio of the 'rise' of the line to the 'run' of the line—the ratio of the vertical change to the horizontal change. More briefly,

$$\text{Slope} = \frac{\text{Rise}}{\text{Run}}$$

We will consider that the 'run' is always to the right. Thus, the 'rise' will be up if the slope is positive and down if the slope is negative. A slope of $\frac{2}{3}$ would indicate that the line rises 2 units for every 3 units of run; a slope of $-\frac{3}{5}$ would indicate that the line 'falls' (a negative 'rise') 3 units for every 5 units of run.

Referring to Fig. 9.3, we see that L_1 has a slope of 1, L_2 has a slope of $-\frac{1}{2}$, and L_3 has a slope of 4. The reader should make sure that he understands the above definitions of slope by verifying the slopes of these three lines.

The equation of the line whose y-intercept is the number a, and whose slope is the number b, is

$$y = bx + a \qquad\qquad (9.7)$$

The converse of the preceding statement is also true: the line whose equation is (9.7) has slope equal to b, and y-intercept equal to a.

Referring to Fig. 9.3, it can be determined that L_1 has the equation $y = x + 1$; L_2 has the equation $y = -\frac{1}{2}x + 1$; and L_3 has the equation $y = 4x - 2$. Again, the reader should verify these statements.

Now that the basic ideas of writing the equations of straight lines have been reviewed, we can return to the problem of finding the line that best expresses the linear relationship between the x- and y-values.

The observations occur in pairs. There are n values of x: $-x_1, x_2, \ldots x_n$; and n values of y: $-y_1, y_2, \ldots y_n$; y_1 corresponds to x_1, y_2 to x_2, and so on. The n pairs of observations can be written $(x_1, y_1), (x_2, y_2), \ldots (x_n, y_n)$.

A scatter diagram of the n pairs of observations is made (see Fig. 9.1). We wish to find the straight line that best fits these n plotted points. A straight line could be fitted to the n points of the scatter diagram merely by eye—that is, the experimenter just estimates where the line of best fit should go, then takes a ruler and draws it. A typical scatter diagram and a line fitted to the points by eye is shown in Fig. 9.4.

We see that the y-intercept is about 38. The value of the slope can be estimated (by eye) as about $\frac{1}{2}$. Therefore the equation of the fitted line is approximately

$$y = \tfrac{1}{2}x + 38$$

So for any particular x-value, say x_i, the corresponding y_i would be given by

$$y_i = \tfrac{1}{2}x_i + 38$$

This method is rather crude and has serious shortcomings, although it is not entirely unsatisfactory. One objection, for instance, is that different persons would get different 'best-fitting' lines. Fortunately, there is a way to get a line that is certainly the best-fitting line, and there is only one such line for any set of n points.

First, we will define the best-fitting line as the line for which the sum of the squares of deviations of the predicted y-values from the observed y-values is a

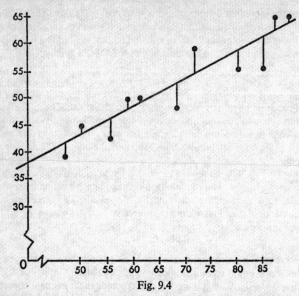

Fig. 9.4

minimum. (In Fig. 9.4 the observed values are the dots, the predicted values are on the line, and the deviations are the vertical lines that connect the observed and predicted values.) Defining the best-fitting line in this manner, it can be shown, by methods that are beyond the scope of this book, that the equation of the best-fitting line is

$$\hat{y} = b \cdot x + (\bar{y} - b \cdot \bar{x}) \tag{9.8}$$

where

$$\bar{x} = \frac{\sum x_i}{n}$$

$$\bar{y} = \frac{\sum y_i}{n}$$

and

$$b = \frac{\sum (x_i - \bar{x})y_i}{\sum (x_i - \bar{x})^2}$$

The quantity b is the slope of the regression line and $(\bar{y} - b \cdot \bar{x})$ is the y-intercept.

The symbol $\hat{y}$ is read 'y hat', and is used to distinguish the predicted y's (on the line in Fig. 9.4) from the observed y's (the dots in Fig. 9.4). Given a particular x-value, say x_j, the predicted y-value, $\hat{y}_j$, is found by substituting the given x-value into Formula (9.8):

$$\hat{y}_j = b \cdot x_j + (\bar{y} - b \cdot \bar{x}) \tag{9.9}$$

The best computing formula for b is

$$b = \frac{\sum xy - \frac{(\sum x)(\sum y)}{n}}{\sum x^2 - \frac{(\sum x)^2}{n}} \tag{9.10}$$

In most problems the quantities $\sum x, \sum y, \sum x^2$, and $\sum xy$ can be found simultaneously on a desk calculator.

EXAMPLE: A certain company would like to predict how the sales trainees in its salesmanship course will perform. At the beginning of their two-month course the trainees are given an aptitude test. This is the x-score shown in Table 9.5. Records are kept of the sales record of each salesman. The y-values in Table 9·7 give the sales (for the year following the training period) of each of the individuals who took the aptitude test. (The amount of sales is given in thousands of £.)

Table 9.5

x	y
18	54
26	64
28	54
34	62
36	68
42	70
48	76
52	66
54	76
60	74

The necessary sums are found (using a desk calculator) to be $\sum x = 398$, $\sum y = 664$, $\sum x^2 = 17\,524$, and $\sum xy = 27\,268$.

Substituting these values into Formula (9.10) we have

$$b = \frac{27\,268 - \dfrac{(398)(664)}{10}}{17\,524 - \dfrac{(398)^2}{10}} = \frac{27\,268 - 26\,427}{17\,524 - 15\,840}$$

$$= \frac{841}{1684} = 0.4994$$

also $\quad \bar{x} = 39.8000$

and $\quad \bar{y} = 66.4000$

Substituting these values into Formula (9.9), we obtain the estimated regression line

$$\hat{y}_j = 0.499x_j + (66.4 - 19.9)$$
$$\hat{y}_j = 0.499x_j + 46.5$$

TESTING HYPOTHESES ABOUT μ IN A REGRESSION PROBLEM

We can test hypotheses about μ and β, the parameters for which $\bar{y}$ and b are estimates, by means of the t-distribution. (The parameter μ is the true mean of the y's, and β is the true slope of the regression line.)

If the null hypothesis H_0: $\mu = \mu_0$ is true the random variable

$$\frac{\bar{y} - \mu_0}{\sigma_{\bar{y}}}$$

where $\sigma_{\bar{y}}^2$ denotes the theoretical variance of the random variable $\bar{y}$, has a standard normal distribution. When the theoretical variance $\sigma_{\bar{y}}^2$ is unknown, it must be estimated by $s_{\bar{y}}^2$. The random variable

$$\frac{\bar{y} - \mu_0}{s_{\bar{y}}}$$

has a t-distribution with $n - 2$ degrees of freedom, if the null hypothesis $H_0: \mu = \mu_0$ is true. The number of degrees of freedom of the t-distribution is the same as the divisor used in finding s^2 (see Formula 9.11), which in its turn is used in finding s_y (see Formula 9.12). The quantity $\bar{y}$ is the mean of a sample of n observations. We know that if y has a variance equal to σ^2, then $\bar{y}$ has a variance equal to $\dfrac{\sigma^2}{n}$. Symbolically, $\sigma_{\bar{y}}^2 = \dfrac{\sigma^2}{n}$.

It can be shown that an estimate of σ^2 is given by

$$s^2 = \frac{\sum (y_i - \hat{y}_i)^2}{n - 2} \tag{9.11}$$

The quantity s^2 is called the **error mean square**. The square root of s^2, s, is called the **standard error of the estimate**. Then, since s^2 is our estimate of σ^2, the estimate of $\sigma_{\bar{y}}^2$ is

$$s_{\bar{y}}^2 = \frac{s^2}{n} \tag{9.12}$$

Examining Formula (9.11), we see that it is necessary to find $\sum (y_i - \hat{y}_i)^2$, where the y_i are the predicted y-values. In order to find s^2, using Formula (9.11), we must first find the predicted y-value for every x-value, then find the differences $y_i - \hat{y}_i$, square these differences, and add them. These computational operations take time and, as often is the case, there is a time-saving computational formula which does not require the computation of the predicted y-values. If the predicted values are going to be calculated whether they are needed in the computation of s^2 or not (the experimenter might want to see just how close the predicted values are to the observed values), then it does not make much difference which method is used. If the predicted values are not going to be found unless they are necessary the quantity $\sum (y_i - \hat{y}_i)^2$ can be found most conveniently from the computing formula

$$\sum (y_i - \hat{y}_i)^2 = \sum (y_i - \bar{y})^2 - b \cdot \sum (x_i - \bar{x})y_i \tag{9.13}$$

Each of the quantities on the right-hand side of Formula (9.13) has a computing formula, too: b is the slope of the regression line and is calculated from Formula (9.10); $\sum (x_i - \bar{x})y_i$ is just the numerator of b, and is found during the computation of b; $\sum (y_i - \bar{y})^2$ is found from the formula

$$\sum (y_i - \bar{y})^2 = \sum y_i^2 - \frac{(\sum y_i)^2}{n} \tag{9.14}$$

The estimate of the population variance is found by each method in the two following examples.

EXAMPLE: Use the data in Table 9.5 to find s^2, finding $\sum (y_i - \hat{y}_i)^2$ without the aid of the computing formula given in (9.13).

If the computations are being done on a desk calculator it is convenient to use a constant multiplier (see the handbook furnished with the calculator for directions on how to do this).

The observed y-values from Table 9.5, along with the predicted y-values, the differences $y_i - \hat{y}_i$, and the squared differences $(y_i - \hat{y}_i)^2$ are shown in Table 9.6.

Table 9.6

y_i	$\hat{y}_i$	$y_i - \hat{y}_i$	$(y_i - \hat{y}_i)^2$
54	55·5	−1·5	2·25
64	59·5	4·5	20·25
54	60·5	−6·5	42·25
62	63·5	−1·5	2·25
68	64·5	3·5	12·25
70	67·5	2·5	6·25
76	70·5	5·5	30·25
66	72·4	−6·4	40·96
76	73·4	2·6	6·76
74	76·4	−2·4	5·76
			$169 \cdot 23 = \sum (y_i - \hat{y}_i)^2$

We see that $\sum (y_i - \hat{y}_i)^2 = 169\cdot23$. Substituting this value into Formula (9.11), we get

$$s^2 = \frac{169\cdot23}{8} = 21\cdot2$$

correct to the nearest tenth.

EXAMPLE: Use the data in Table 9.5 to find s^2, finding $\sum (y_i - \hat{y}_i)^2$ by the computing formula in (9.13).

We need five sums—$\sum x, \sum y, \sum x^2, \sum y^2$, and $\sum xy$. All of these except $\sum y^2$ were found in the preceding section. Calculating $\sum y_i^2$ and copying the others from that section, we have $\sum x = 398$, $\sum y = 664$, $\sum x^2 = 17\,524$, $\sum y^2 = 44\,680$, and $\sum xy = 27\,268$.

The value of b has already been found to be 0·4994. The quantity $\sum (x - \bar{x})y$ is the numerator of b, and has already been found to be 841. The third necessary quantity is

$$\sum (y - \bar{y})^2 = \sum y^2 - \frac{\left(\sum y\right)^2}{n} = 44\,680 - \frac{(664)^2}{10} = 590$$

substituting these values into Formula (9.13), we have

$$\sum (y_i - \hat{y}_i)^2 = 590 - (0\cdot4994)(841)$$
$$= 590 - 420$$
$$= 170$$

correct to the nearest integer.

Then, substituting this value of $\sum (y_i - \hat{y}_i)^2$ into Formula (9.11), we have

$$s^2 = \frac{170}{8} = 21\cdot3$$

correct to the nearest tenth, which agrees closely with the value of 21·2 found by the first procedure. The discrepancy of 0·1 is due to rounding error.

Procedure for Testing H_0: $\mu = \mu_0$ versus H_1: $\mu \neq \mu_0$

(1) Formulate the null and alternative hypotheses: H_0: $\mu = \mu_0$ versus H_1: $\mu \neq \mu_0$.

(2) Decide on an α-level.

Find $t_{\alpha/2}(n - 2)$ in the t-table.

The critical region consists of two parts: $t \leqslant -t_{\alpha/2}(n - 2)$ and $t \geqslant t_{\alpha/2}(n - 2)$.

(3) Select a random sample; measure or otherwise determine x and y.

Find the regression line (see Formula (9.8) and following).

Find $s_{\bar{y}}$ (see Formula (9.11) and following).

(4) Calculate $t = \dfrac{\bar{y} - \mu_0}{s_{\bar{y}}}$

(5) If t is in the critical region, reject H_0.

When the alternative hypothesis is H_1: $\mu < \mu_0$ the procedure is the same as outlined above, with one exception. Step (2) would be

(2) Decide on an α-level.

Find $t_\alpha(n - 2)$ in the t-table.

The critical region is $t \leqslant -t_\alpha(n - 2)$.

When the alternative hypothesis is H_1: $\mu > \mu_0$, step (2) would be

(2) Decide on an α-level.

Find $t_\alpha(n - 2)$ in the t-table.

The critical region is $t \geqslant t_\alpha(n - 2)$.

EXAMPLE: Using the data given in Table 9.5, which yielded the regression line $\hat{y}_i = 0 \cdot 499 x_i + 46 \cdot 5$, test the hypothesis that $\mu = 55$ versus the alternative that μ is greater than 55.

(1) We are to test H_0: $\mu = 55$ versus H_1: $\mu > 55$.

(2) Let $\alpha = 0 \cdot 05$.

The number of degrees of freedom is $n - 2 = 8$, and $t_{0 \cdot 05}(8) = 1 \cdot 86$.

The critical region consists of all values of t greater than or equal to $1 \cdot 86$.

(3) The random sample has already been given, in Table 9.5.

The regression line has been found to be $\hat{y}_i = 0 \cdot 499 x_i + 46 \cdot 5$.

The value of s^2 is $21 \cdot 3$, from the preceding example in this section. We thus have

$$s_{\bar{y}}^2 = \frac{s^2}{n} = \frac{21 \cdot 3}{10} = 2 \cdot 13$$

and

$$s_{\bar{y}} = 1 \cdot 46$$

(4)

$$\bar{y} = \frac{\sum y}{n} = \frac{664}{10} = 66 \cdot 4$$

$$t = \frac{66 \cdot 4 - 55}{1 \cdot 46} = \frac{11 \cdot 4}{1 \cdot 46} = 7 \cdot 81$$

(5)

$$7 \cdot 81 > 1 \cdot 86$$

Reject H_0 and accept H_1.

TESTING HYPOTHESES ABOUT β IN A REGRESSION PROBLEM

In the last section we discussed the testing of hypotheses about μ. The procedure for testing a hypothesis about β is analogous.

If the null hypothesis H_0: $\beta = \beta_0$ is true, and if the theoretical variance of the random variable b is known, then the random variable

$$\frac{\mathbf{b} - \beta_0}{\sigma_b}$$

has a standard normal distribution. If σ_b is unknown and must be estimated by s_b the random variable

$$\frac{\mathbf{b} - \beta_0}{s_b}$$

has a t-distribution with $n - 2$ degrees of freedom.

The estimated variance of the random variable $\mathbf{b}$ is more complicated than the variance of the random variable $\bar{\mathbf{y}}$, although it, too, makes use of the sum of squares of error. The estimated variance of $\mathbf{b}$ is

$$s_b^2 = \frac{s^2}{\sum(x_i - \bar{x})^2} \tag{9.15}$$

where s^2 is given by Formula (9.11).

Procedure for Testing H_0: $\beta = \beta_0$ versus H_1: $\beta \neq \beta_0$

(1) Formulate the null and alternative hypotheses: H_0: $\beta = \beta_0$ and H_1: $\beta \neq \beta_0$.

(2) Decide on an α-level.

Find $t_{\alpha/2}(n - 2)$ in the t-table.

The critical region consists of two parts: $t \leqslant -t_{\alpha/2}(n - 2)$ and $t \geqslant t_{\alpha/2}(n - 2)$.

(3) Select a random sample and measure (or otherwise determine) the x and y values.

Find the regression line (Formula (9.8)).

Calculate s^2 (Formula (9.11) and following).

Calculate s_b^2 from Formula (9.15).

(4) Calculate $t = \dfrac{b - \beta_0}{s_b}$

(5) Note whether the value of t is in the critical region, and accept or reject H_0.

The steps in the above procedure for testing the null hypothesis H_0: $\beta = \beta_0$ versus a two-sided alternative remain nearly the same when a one-sided alternative is tested instead.

If the one-sided alternative H_1: $\beta > \beta_0$ is tested, step (2) should be changed to read as follows:

(2) Decide on an α-level.

Find $t_{\alpha}(n - 2)$ in the t-table.

The critical region is $t \geqslant t_{\alpha}(n - 2)$.

If the one-sided alternative $H_1: \beta < \beta_0$ is tested, step (2) should be changed to read as follows:

(2) Decide on an α-level.
   Find $t_\alpha(n - 2)$ in the t-table.
   The critical region is $t \leqslant - t_\alpha(n - 2)$.

The most frequently tested hypothesis is probably $H_0: \beta = 0$. If there is a linear relationship between x and y, then β will have a non-zero value. (If β equalled zero, this would indicate that the mean of y is the same for any value of x, which implies that there is no relationship between x and y.) Therefore, testing the hypothesis that β equals zero is equivalent to testing the hypothesis that there is no linear relationship between x and y.

EXAMPLE: Use the data in Table 9.5 to test the hypothesis that there is no linear relationship between x and y, versus the alternative that there is.

(1) We are to test $H_0: \beta = 0$ versus $H_1: \beta \neq 0$.
(2) Let $\alpha = 0 \cdot 01$.
   The sample size is 10, and the number of degrees of freedom is 8.
   $t_{0 \cdot 005}(8) = 3 \cdot 355$.
(3) The sample is given.
   The regression line has been found to be $\hat{y}_j = 0 \cdot 499 x_j + 46 \cdot 5$.
   We have calculated

$$s^2 = 21 \cdot 3 \quad \text{(second example in the preceding section)}$$

and $\sum (x - \bar{x})^2 = 1684$ (the denominator of b, found earlier).

Substituting these two quantities into Formula (9.15), we have

$$s_b^2 = \frac{21 \cdot 3}{1684} = 0 \cdot 0126$$

and $$s_b = 0 \cdot 112$$

(4) We have previously calculated

$$b = 0 \cdot 4994$$
$$t = \frac{0 \cdot 4994 - 0}{0 \cdot 112} = 4 \cdot 46$$

(5) t is in the critical region. Reject H_0 and accept H_1.

EXERCISES

1. Given that a random sample of size 8 yielded the following data, find the sample correlation coefficient.

x	y
25	16
18	11
32	20
27	17
21	15
35	26
28	32
30	20

2. Find the sample correlation coefficient for the following data. (If you wish, you can subtract a convenient number from each x-value and a convenient number from each y-value before performing the computation.)

x	y
380	560
402	543
370	564
365	573
410	550
392	554
385	540

3. A sample of size 28 yielded $r = -0.56$. Test $H_o: \rho = -0.45$ against the two-sided alternative. Let $\alpha = 0.05$.

4. A sample of size 12 yielded $r = 0.32$. Test $H_o: \rho = 0$ versus $H_1: \rho \neq 0$. Let $\alpha = 0.01$.

5. A sample of size 19 yielded $r = 0.60$. Test $H_o: \rho = 0.7$ against the alternative, $H_1: \rho > 0.7$.

6. Find the least-squares line for the data of Exercise 1.

7. Find the standard error of the estimate for the least-squares line found in Exercise 6.

8. For the least-squares line in Exercise 6, test $H_o: \beta = 0$ against the alternative hypothesis, $H_1: \beta > 0$.

9. For the least-squares line in Exercise 6, test $H_o: \beta = 1$ against the alternative hypothesis, $H_1: \beta \neq 1$.

10. For the least-squares line in Exercise 6, test $H_o: \mu = 18$ against the alternative hypothesis, $H_1: \mu > 18$.

11. Given the data below, find the regression (least-squares) line:

x	y
20	5
11	15
15	14
10	17
17	8
19	9

12. For the regression line found in Exercise 11, find the predicted values of y for $x = 10, 11, 15, 17, 19$, and 20.

13. Use the predicted y-values found in Exercise 12 to find the standard error of the estimate for the regression line that was found in Exercise 11.

14. For the regression line found in Exercise 11, test $H_o: \beta = 0$ versus $H_1: \beta < 0$. Let $\alpha = 0.01$.

CHAPTER X

CONFIDENCE LIMITS

INTRODUCTION

Previously we have used sample data to provide estimates of population parameters. For example, if we find $\bar{x} = 50$ for a certain sample we use 50 as our estimate of μ. We have no reason to believe, however, that μ equals 50 exactly; we merely hope that μ is close to 50. Or if we find that $s^2 = 100$ for a certain sample we use 100 as our estimate of σ^2. Once again we have no reason to believe that σ^2 equals exactly 100, but we expect the value of σ^2 to be in the vicinity of 100. Estimates such as $\bar{x} = 50$ and $s^2 = 100$ are known as **point estimates** because they are single numbers.

It is frequently desirable to know that a parameter is likely to be in a certain *interval* of values. For example, if we obtain $\bar{x} = 50$ for a certain sample, perhaps it is likely that μ lies in the interval from 45 to 55. Or if we obtain $s^2 = 100$ in a certain sample it may be likely that σ^2 lies in the interval from 87 to 129. Estimates such as these are known as **interval estimates**, because an interval within which the parameter might reasonably be found is given. Statements such as the above are not very meaningful until we know *how* likely it is that μ lies in a certain interval, or that σ^2 lies in a certain interval. Therefore, a number known as a **confidence coefficient** or as a **degree of confidence** is given along with the interval, which is called a **confidence interval**. We will use the term 'degree of confidence', rather than 'confidence coefficient', although the two terms are synonymous and are used interchangeably.

A convenient notation for the interval from the number a to the number b is (a, b). Thus, (45, 55) represents the interval from 45 to 55, and (87, 129) represents the interval from 87 to 129. The pairs of numbers, 45 and 55, and 87 and 129, are known as **confidence bounds** or **confidence limits**. The words 'bound' and 'limit' have exactly the same meaning in this chapter, although 'limit' will be used henceforth. The smaller of the two confidence limits is called the **lower confidence limit**, and the larger is called the **upper confidence limit**. Thus, for the confidence interval (45, 55), 45 is the lower confidence limit and 55 is the upper confidence limit. Similarly, for the confidence interval (87, 129), 87 is the lower confidence limit and 129 is the upper confidence limit. When there is no danger of misunderstanding the word 'confidence' is often omitted, and the numbers 87 and 129 are simply referred to as the lower limit and the upper limit, respectively.

Usually a confidence statement has the following form: 'A 90 per cent [for instance] confidence interval for μ is (45, 55).' This means that we are 90 per cent confident that μ lies in the interval (45, 55). The confidence interval is (45, 55) and the confidence coefficient is 0·90. The statement 'We are 90 per cent confident that μ lies in the interval (45, 55)' does *not* mean that the *probability* that μ lies in the interval (45, 55) is 0·90. It means that if we would draw a large number of samples from the population and find a confidence interval for μ for each sample, then about 90 per cent of the intervals would contain μ. Each of the large number of confidence intervals is found by using the appropriate formula in the section where confidence intervals for μ are discussed.

This distinction between a confidence statement and a probability statement will be discussed at more length later in this chapter.

A Note on Inequalities. The same number (positive or negative) can be added to both sides of an inequality without changing the direction of the inequality sign. Stated mathematically, if $x < y$, then $x + a < y + a$, where a is any number. (Note that the 'point' of the inequality symbol is towards the left in both cases.) Specifically, if $x < 3$, then $x + 2 < 5$ (2 has been added to both sides); if $x + 4 > 7$, then $x > 3$ (−4 has been added to both sides).

Both sides of an inequality can be multiplied by the same positive number without changing the direction of the inequality symbol. If $x < y$, then $bx < by$, where b is a positive number. If $x < y$, then $2x < 2y$ (both sides have been multiplied by 2); if $3x > 5$, then $x > \frac{5}{3}$ (both sides have been multiplied by $\frac{1}{3}$).

When both sides of an equality are multiplied by the same negative number the direction of the inequality symbol must be reversed. If $x < y$, then $cx > cy$, where c is a negative number. If $-x > 2$, then $x < -2$ (both sides have been multiplied by -1); if $-4x < 3$, then $x > -\frac{3}{4}$ (both sides have been multiplied by $-\frac{1}{4}$).

Frequently one meets with inequalities of the form

$$3 < 2x < 5$$
or
$$-5 < -x < 2$$

We will call inequalities such as these **continued inequalities.** Each of the quantities 3, or $2x$, or 5 will be called **members** of the continued inequality. The first is read '3 is less than $2x$ and $2x$ is less than 5' and simply means that the value of $2x$ is between 3 and 5. In order to solve the first inequality for x, we must manipulate the inequality until the x is alone between the two inequality symbols. We concentrate on changing the $2x$ into x, and whatever we do to the $2x$, we must do to the other members of the inequality, also. In order to change $2x$ into x we must multiply the $2x$ by $\frac{1}{2}$. Having multiplied the centre members by $\frac{1}{2}$, we must multiply the other two members by $\frac{1}{2}$. Thus the first inequality becomes

$$\frac{3}{2} < x < \frac{5}{2}$$

Thus x is between $\frac{3}{2}$ and $\frac{5}{2}$—the inequality has been solved for x.

To solve the second of the above inequalities for x, we must multiply the $-x$ in the centre by -1 to obtain an x in the centre. Whatever is done to one member of an inequality must be done to the others. We must multiply the -5 and the 2 by -1. Keep in mind that when we multiply inequalities by negative numbers the direction of the inequality symbols must be reversed. The second inequality becomes, after multiplying all members by -1 and reversing the direction of the inequality symbols,

$$5 > x > -2$$

It is preferable to write the smaller number to the left and the larger to the right. The inequality immediately above can be written

$$-2 < x < 5$$

EXAMPLE: Solve $-5 < -4x + 3 < -1$ for x.

Adding -3 to each term, the inequality becomes

$$-8 < -4x < -4$$

Multiplying each term by $-\frac{1}{4}$, we obtain

$$2 > x > 1$$

which is equivalent to $\qquad 1 < x < 2$

CONFIDENCE INTERVALS FOR μ

Recall (in Chapters VII and VIII) that several different cases were treated when we discussed testing hypotheses about the mean of a population: (1) the population is normal with known variance; (2) the population is normal with unknown variance; and (3) the population is not normal, and the sample size is large (at least 30). In this section we will also find the expression that yields confidence intervals for the population mean for each of these cases.

(1) *Normal Population, Known Variance*

If we are sampling from a normal population with unknown mean μ, and known variance σ^2, then the random variable $\bar{x}$ (based on a sample of size n) has a normal distribution with mean μ and variance σ^2/n. Therefore the random variable $\dfrac{\bar{x} - \mu}{\sigma/\sqrt{n}}$ has a standard normal distribution and, knowing this fact, we can say that

$$P\left\{-z_{\alpha/2} < \frac{\bar{x} - \mu}{\sigma/\sqrt{n}} < z_{\alpha/2}\right\} = 1 - \alpha \qquad (10.1)$$

where, as usual, $z_{\alpha/2}$ represents the upper $\alpha/2$ point of the standard normal distribution.

We want to 'solve' the inequality within the brackets for the parameter μ. Multiplying by $\sigma/\sqrt{n}$, the inequality becomes

$$-z_{\alpha/2} \cdot (\sigma/\sqrt{n}) < \bar{x} - \mu < z_{\alpha/2} \cdot (\sigma/\sqrt{n}) \qquad (10.2)$$

Now we see that we must eliminate the $\bar{x}$ in the centre if we wish to have the μ alone. Adding $-\bar{x}$ to all terms (or, equivalently, subtracting $\bar{x}$ from each term), (10.2) becomes

$$-\bar{x} - z_{\alpha/2} \cdot (\sigma/\sqrt{n}) < -\mu < -\bar{x} + z_{\alpha/2} \cdot (\sigma/\sqrt{n}) \qquad (10.3)$$

Multiplying each term of (10.3) by -1, we obtain (remember: the direction of the inequality signs must be reversed)

$$\bar{x} + z_{\alpha/2} \cdot (\sigma/\sqrt{n}) > \mu > \bar{x} - z_{\alpha/2} \cdot (\sigma/\sqrt{n}) \qquad (10.4)$$

which can be rewritten as

$$\bar{x} - z_{\alpha/2} \cdot (\sigma/\sqrt{n}) < \mu < \bar{x} + z_{\alpha/2} \cdot (\sigma/\sqrt{n}) \qquad (10.5)$$

The inequality in Formula (10.5) is equivalent to the inequality within the braces in (10.1); therefore (10.1) can be written

$$P\{\bar{x} - z_{\alpha/2} \cdot (\sigma/\sqrt{n}) < \mu < \bar{x} + z_{\alpha} \cdot (\sigma/\sqrt{n})\} = 1 - \alpha \qquad (10.6)$$

A $100(1 - \alpha)$ per cent confidence interval for μ for a particular sample is

$$(\bar{x} - z_{\alpha/2} \cdot \sigma/\sqrt{n}, \quad \bar{x} + z_{\alpha/2} \cdot \sigma/\sqrt{n}) \qquad (10.7)$$

Formula (10.7) yields an exact $100(1 - \alpha)$ per cent confidence interval for μ, assuming that we are sampling from a normal population with known variance. Note especially that the quantities $\bar{x} - z_{\alpha/2} \cdot \sigma/\sqrt{n}$ and $\bar{x} + z_{\alpha/2} \cdot \sigma/\sqrt{n}$ in (10.7) are numerical values, not random variables. For the particular sample, $\bar{x}$ (a particular value of $\bar{\mathbf{x}}$) is calculated, then the lower and upper confidence limits are found.

The difference between a *probability statement* and a *confidence statement* must be clearly understood. A probability statement is made about random variables. For instance, the symbolic statement in Formula (10.6) is a correct statement because $\bar{\mathbf{x}}$ is a random variable. The statement represents *potentiality* —$\bar{\mathbf{x}}$ does not have any particular numerical value, but takes on different numerical values in different samples. After a sample is drawn, then a *numerical value* of the random variable $\bar{\mathbf{x}}$ can be found; this numerical value is denoted $\bar{x}$. The reader should distinguish carefully between the random variable, $\bar{\mathbf{x}}$, and the numerical value of the random variable for a particular sample, $\bar{x}$: $\bar{x}$ is a numerical value of $\bar{\mathbf{x}}$. The probability that the random variable $\bar{\mathbf{x}}$ will take on a value $\bar{x}$ such that μ will be between the numbers $\bar{x} - z_{\alpha/2} \cdot \sigma/\sqrt{n}$ and $\bar{x} + z_{\alpha/2} \cdot \sigma/\sqrt{n}$ is $1 - \alpha$. After the random variable $\bar{\mathbf{x}}$ has been replaced by a particular numerical value the lower confidence limit is the *number* $\bar{x} - z_{\alpha/2} \cdot \sigma/\sqrt{n}$ and the upper confidence limit is the *number* $\bar{x} + z_{\alpha/2} \cdot \sigma/\sqrt{n}$ It is incorrect to say that

$$P\{\bar{x} - z_{\alpha/2} \cdot \sigma/\sqrt{n} < \mu < \bar{x} + z_{\alpha/2} \cdot \sigma/\sqrt{n}\} = 1 - \alpha$$

because the quantity within the braces is not a random variable.

The statement is either true or false after the numerical values of the confidence limits have been calculated. If it is true, then the probability is 1; if it is false the probability is 0. In either case it is meaningless to say that the probability is $1 - \alpha$.

For a particular sample the confidence interval either contains μ or it does not. For example, suppose that $(53\cdot5, 67\cdot0)$ is a 95 per cent confidence interval for μ. It is erroneous to say $P(53\cdot5 < \mu < 67\cdot0) = 0\cdot95$. The true value of μ is unknown, of course, but if μ lies within the interval, then

$$P(53\cdot5 < \mu < 67\cdot0) = 1$$

To be specific, if the value of μ is $65\cdot2$, say, then the probability that μ is contained in the interval $(53\cdot5, 67\cdot0)$ is 1; that is,

$$P\{53\cdot5 < 65\cdot2 < 67\cdot0\} = 1$$

On the other hand, if μ lies outside the interval, then

$$P\{53\cdot5 < \mu < 67\cdot0\} = 0$$

Specifically, if $\mu = 71\cdot3$, say, then

$$P\{53\cdot5 < 71\cdot3 < 67\cdot0\} = 0$$

In neither case does $P\{53\cdot5 < \mu < 67\cdot0\}$ equal $0\cdot95$.

The above discussion illustrates the statement that it is incorrect to say that

$$P\{\bar{x} - z_{\alpha/2} \cdot \sigma/\sqrt{n} < \mu < \bar{x} + z_{\alpha/2} \cdot \sigma/\sqrt{n}\} = 1 - \alpha$$

Instead, we say that

$$(\bar{x} - z_{\alpha2/} \cdot \sigma/\sqrt{n}, \quad \bar{x} + z_{\alpha/2} \cdot \sigma/\sqrt{n})$$

is a $100(1 - \alpha)$ per cent confidence interval for μ. This statement means that if we draw a large number of samples (from a normal population with known variance) and compute the numbers

$$\mu_L = \bar{x} - z_{\alpha/2} \cdot \sigma/\sqrt{n}$$

and
$$\mu_U = \bar{x} + z_{\alpha/2} \cdot \sigma/\sqrt{n}$$

for each sample, then about $100(1 - \alpha)$ per cent of the time the interval (μ_L, μ_U) will contain μ.

Thus, instead of saying that the probability is $1 - \alpha$ that μ lies between μ_L and μ_U, we say that we are $100(1 - \alpha)$ per cent confident that μ lies between μ_L and μ_U, and this means that about $100(1 - \alpha)$ per cent of the intervals found in this manner will contain μ.

Procedure: Confidence Interval for μ (Variance Known)

(1) Decide upon the confidence coefficient, $1 - \alpha$: Find $z_{\alpha/2}$ in standard normal tables.

(2) Obtain a random sample of size n.

(3) Calculate $\bar{x}$.
　　Calculate $\sigma/\sqrt{n}$.

(4) Calculate $\mu_L = \bar{x} - z_{\alpha/2} \cdot \sigma/\sqrt{n}$
　　and　　$\mu_U = \bar{x} + z_{\alpha/2} \cdot \sigma/\sqrt{n}$.

(5) A $100(1 - \alpha)$ per cent confidence interval is (μ_L, μ_U).

EXAMPLE: A sample of size 16 from a normal population with known variance $\sigma^2 = 25$ yielded a sample mean equal to 14·5. Within what interval can we be 90 per cent confident that μ lies?

(1) The confidence coefficient is 0·90; therefore, $\alpha = 0 \cdot 10$. $z_{0.05} = 1 \cdot 65$ (correct to 2 decimal places).

(2) Unnecessary in this problem.

(3) $\bar{x} = 14 \cdot 5$ (given).

$$\sigma/\sqrt{n} = 5/\sqrt{16} = \tfrac{5}{4} = 1 \cdot 25$$

(4)
$$\mu_L = 14 \cdot 5 - (1 \cdot 65)(1 \cdot 25) = 12 \cdot 4$$
$$\mu_U = 14 \cdot 5 + (1 \cdot 65)(1 \cdot 25) = 16 \cdot 6$$

(5) The 90 per cent confidence interval is $(12 \cdot 4, 16 \cdot 6)$.

(2) *Normal Distribution, Variance Unknown*

When we are sampling from a normal distribution with unknown variance, the random variable $\dfrac{\bar{x} - \mu}{s/\sqrt{n}}$ has a t-distribution with $n - 1$ degrees of freedom. Therefore, we can say that

$$P\left\{-t_{\alpha/2} < \frac{\bar{x} - \mu}{s/\sqrt{n}} < t_{\alpha/2}\right\} = 1 - \alpha \qquad (10.8)$$

Confidence Limits 155

After performing the same sequence of steps as in Case (1), we can show that the probability statement in Formula (10.8) is equivalent to

$$P\left\{\bar{x} - t_{\alpha/2}\frac{s}{\sqrt{n}} < \mu < \bar{x} + t_{\alpha/2}\frac{s}{\sqrt{n}}\right\} = 1 - \alpha \qquad (10.9)$$

An exact $100(1 - \alpha)$ per cent confidence interval for μ is given by

$$(\bar{x} - t_{\alpha/2} \cdot s/\sqrt{n}, \quad \bar{x} + t_{\alpha/2} \cdot s/\sqrt{n}) \qquad (10.10)$$

Procedure: Confidence Interval for μ (Variance Unknown)

(1) Decide upon the degree of confidence, $1 - \alpha$.
   Decide upon the sample size, n.
   Find $t_{\alpha/2}(n - 1)$ in the t-table.
(2) Obtain a random sample of size n.
   Calculate $\bar{x}$.
   Calculate s.
   Calculate $s/\sqrt{n}$.
(3) Calculate $\mu_L = \bar{x} - t_{\alpha/2} \cdot s/\sqrt{n}$ and $\mu_U = \bar{x} + t_{\alpha/2} \cdot s/\sqrt{n}$.
(4) The $100(1 - \alpha)$ per cent confidence interval is (μ_L, μ_U).

EXAMPLE: A sample of size 16 from a normal population with unknown variance yielded the sample values $\bar{x} = 14.5$ and $s^2 = 25$. Find a 90 per cent confidence interval on the population mean.

(1) The degree of confidence is 0.90. The sample size is 16.

$$t_{\alpha/2}(n - 1) = t_{0.05}(15) = 1.753$$

(2) The sample has already been obtained.

$$\bar{x} = 14.5$$
$$s = \sqrt{25} = 5$$
$$s/\sqrt{n} = 5/\sqrt{16} = 5/4 = 1.25$$

(3) $\qquad \mu_L = 14.5 - (1.753)(1.25) = 14.5 - 2.2 = 12.3$
$\qquad \mu_U = 14.5 + (1.753)(1.25) = 14.5 + 2.2 = 16.7$

(4) The 90 per cent confidence interval is (12.3, 16.7). Note that the data given in this example are exactly the same as in the preceding example except that the population variance is *estimated* to be 25 in this example, whereas it is *known* to be 25 in the preceding one. The necessity of estimating the population variance means that we must use the t-distribution, which yields a confidence interval for the same data that is just a little wider than the one obtained with the normal distribution.

(3) *Non-Normal Distribution, Variance Known or Unknown*

It is known that, for large samples (more than thirty), the sample mean has a distribution that is approximately normal, and the approximation improves as the sample size increases. If the random variable $\bar{x}$ is approximately normal with mean μ and variance σ^2, then the random variable $\frac{\bar{x} - \mu}{\sigma/\sqrt{n}}$ is approximately standard normal. Therefore, we can say that

$$P\left\{-z_{\alpha/2} < \frac{\bar{x} - \mu}{\sigma/\sqrt{n}} < z_{\alpha/2}\right\} \cong 1 - \alpha \qquad (10.11)$$

The symbol $\cong$ means 'approximately equal to'. Formula (10.11) is exactly the same as Formula (10.1), except that the equality is *exact* in (10.1) and only *approximate* in (10.11). Therefore, after we perform the same sequence of steps on the inequality in (10.11) as we did on the inequality in (10.1) we obtain the *approximate* probability statement

$$P\{\bar{x} - z_{\alpha/2} \cdot \sigma/\sqrt{n} < \mu < \bar{x} + z_{\alpha/2} \cdot \sigma\sqrt{n}\} \cong 1 - \alpha \quad (10.12)$$

An approximate $100(1 - \alpha)$ per cent confidence interval on μ, where μ is the mean of a non-normal population with known variance, is given by

$$(\bar{x} - z_{\alpha/2} \cdot \sigma/\sqrt{n}, \quad \bar{x} + z_{\alpha/2} \cdot \sigma/\sqrt{n}) \quad (10.13)$$

where $\bar{x}$ is based on a sample of at least thirty observations.

If the variance of the population is unknown, it should be estimated by the sample variance, s^2, and σ replaced by s in (10.13). (For large sample sizes, s is usually close to the true value of σ.) An *approximate* $100(1 - \alpha)$ per cent confidence interval for μ, where μ is the mean of a non-normal population with unknown variance, is given by

$$(\bar{x} - z_{\alpha/2} \cdot s/\sqrt{n}, \quad \bar{x} + z_{\alpha/2} \cdot s/\sqrt{n}) \quad (10.14)$$

where $\bar{x}$ is based on a sample of at least thirty observations.

Procedure: Confidence Interval for μ, Non-normal Population

The procedure in this case is exactly the same as in Case 1 with one exception: when σ^2 is unknown s must be calculated and used in place of σ in steps (3) and (4).

EXAMPLE: A sample of 64 observations from a population known to be non-normal yielded the sample values, $\bar{x} = 172$ and $s^2 = 299$. Find an approximate 99 per cent confidence interval for μ.

(1) The confidence coefficient is $0\cdot99$: $z_{0.005} = 2\cdot58$.
(2) The sample has already been obtained.
(3)
$$\bar{x} = 172$$
$$s = \sqrt{299} = 17\cdot29$$
$$s/\sqrt{n} = \frac{17\cdot29}{\sqrt{64}} = \frac{17\cdot29}{8} = 2\cdot16$$
(4)
$$\mu_L = 172 - (2\cdot58)(2\cdot16) = 172 - 5\cdot6 = 166\cdot4$$
$$\mu_U = 172 + (2\cdot58)(2\cdot16) = 172 + 5\cdot6 = 177\cdot6$$

(5) An approximate 99 per cent confidence interval for μ is $(166\cdot4, 177\cdot6)$.

CONFIDENCE INTERVAL FOR π

Let the random variable x denote the number of successes in n binomial trials (equivalently, the number of successes in a random sample of size n obtained from a binomial population). The random variable p, where

$$p = \frac{x}{n}$$

will denote the proportion of successes in n binomial trials, and is an estimator of the parameter π, the proportion of successes in the population. We know that **p** has a distribution that is approximately normal with mean π and variance $\dfrac{\pi(1-\pi)}{n}$ (see Chapter VII). Therefore, the standardized random variable

$$\frac{\mathbf{p}-\pi}{\sigma_p}$$

where
$$\sigma_p = \sqrt{\frac{\pi(1-\pi)}{n}} \qquad (10.15)$$

has a distribution that is approximately standard normal. Consequently, we can make the approximate probability statement

$$P\left\{-z_{\alpha/2} < \frac{\mathbf{p}-\pi}{\sigma_p} < z_{\alpha/2}\right\} \cong 1-\alpha \qquad (10.16)$$

We must manipulate the continued inequality in (10.16) so that the middle member will consist of π only. Multiply each member of the inequality within braces by σ_p, and that inequality becomes

$$-z_{\alpha/2} \cdot \sigma_p < \mathbf{p} - \pi < z_{\alpha/2} \cdot \sigma_p \qquad (10.17)$$

After subtracting **p** from each member of (10.17) we obtain

$$-\mathbf{p} - z_{\alpha/2} \cdot \sigma_p < -\pi < -\mathbf{p} + z_{\alpha/2} \cdot \sigma_p \qquad (10.18)$$

which is equivalent to

$$\mathbf{p} - z_{\alpha/2} \cdot \sigma_p < \pi < \mathbf{p} + z_{\alpha/2} \cdot \sigma_p \qquad (10.19)$$

When we replace the inequality in (10.16) by the equivalent inequality in (10.19) we have the approximate probability statement

$$P\{\mathbf{p} - z_{\alpha/2} \cdot \sigma_p < \pi < \mathbf{p} + z_{\alpha/2} \cdot \sigma_p\} \cong 1-\alpha \qquad (10.20)$$

For a particular sample we observe a number of successes, x, and the sample proportion, denoted $\hat{p}$, is

$$\hat{p} = \frac{x}{n}$$

The quantity $\hat{p}$ is a particular value of the random variable **p**. A $100(1-\alpha)$ per cent confidence interval for π, calculated from a particular sample, is given by

$$(\hat{p} - z_{\alpha/2} \cdot \sigma_p, \quad \hat{p} + z_{\alpha/2} \cdot \sigma_p) \qquad (10.21)$$

As the reader will recall (see (10.15)) the variance of **p** depends upon the value of π, which is unknown. However, the sample proportion, $\hat{p}$, can be used as an estimate of the population proportion, π. (It seems intuitively reasonable to estimate the population proportion by the sample proportion.) Therefore, we can obtain an estimate of the true variance, which we will denote s_p, if we replace π by $\hat{p}$ in Formula (10.15):

$$s_p = \sqrt{\frac{\hat{p}(1-\hat{p})}{n}} \qquad (10.22)$$

When we replace σ_p by s_p, (10.21) becomes

$$(\hat{p} - z_{\alpha/2} \cdot s_p, \quad \hat{p} + z_{\alpha/2} \cdot s_p) \tag{10.23}$$

We will use (10.23) to find an approximate $100(1 - \alpha)$ per cent confidence interval for π.

Procedure: Confidence Interval for π

(1) Decide upon the degree of confidence, $1 - \alpha$.
   Look up $z_{\alpha/2}$ in standard normal tables.
(2) Obtain a random sample of size n.
   Note the number of 'successes', x.
(3) Calculate $\hat{p} = \dfrac{x}{n}$

   Calculate $s_p = \sqrt{\dfrac{\hat{p}(1 - \hat{p})}{n}}$
(4) Calculate $\pi_L = \hat{p} - z_{\alpha/2} \cdot s_p$;
   Calculate $\pi_U = \hat{p} + z_{\alpha/2} \cdot s_p$.
(5) A $100(1 - \alpha)$ per cent confidence interval for π is (π_L, π_U).

EXAMPLE: A random sample of 300 car owners in a certain large city were asked whether their cars were equipped with seat belts. If 60 of the cars were so equipped, find an approximate 95 per cent confidence interval on the true proportion, π, of cars in the city that are equipped with seat belts.

(1) The degree of confidence is 0.95. $z_{0.025} = 1.96$.
(2) The sample has already been obtained.

$$n = 300, \, x = 60$$

(3) $$\hat{p} = \frac{60}{300} = 0.20$$

$$s_p = \sqrt{\frac{(0.20)(0.80)}{300}} = \sqrt{0.000533} = 0.023$$

(4) $$\pi_L = 0.20 - (1.96)(0.023) = 0.20 - 0.045 = 0.155$$
$$\pi_U = 0.20 + (1.96)(0.023) = 0.20 + 0.045 = 0.245$$

(5) An approximate 95 per cent confidence interval for π is $(0.155, 0.245)$

CONFIDENCE INTERVAL FOR $\mu_1 - \mu_2$

The various situations for which we will discuss finding a confidence interval for $\mu_1 - \mu_2$ belong to one of three cases: (1) the two populations are both normal with known variances; (2) the two populations are normal, with equal unknown variances; (3) the populations are not normal, in which case large sample sizes are necessary.

(1) *Normal Populations, Known Variances*

Suppose that the random variable $\bar{x}_1$ is based on a random sample of size n_1 from a normal population with mean μ_1 and known variance σ_1^2. Also, suppose that the random variable $\bar{x}_2$ is based on a sample of size n_2 from a normal population with mean μ_2 and known variance σ_2^2. Then the random variable $\bar{x}_1 - \bar{x}_2$ has a normal distribution with mean $\mu_1 - \mu_2$ and known variance

$$\sigma^2 = \frac{\sigma_1^2}{n_1} + \frac{\sigma_2^2}{n_2} \qquad (10.24)$$

which will be denoted simply by σ^2 hereafter in Case 1 *only*

Therefore, the standardized random variable

$$\frac{(\bar{x}_1 - \bar{x}_2) - (\mu_1 - \mu_2)}{\sigma}$$

has a standard normal distribution, and we can say that

$$P\left\{-z_{\alpha/2} < \frac{(\bar{x}_1 - \bar{x}_2) - (\mu_1 - \mu_2)}{\sigma} < z_{\alpha/2}\right\} = 1 - \alpha \qquad (10.25)$$

When the variances of two normal populations are known, an exact $100(1 - \alpha)$ per cent confidence interval for $\mu_1 - \mu_2$ is given by

$$((\bar{x}_1 - \bar{x}_2) - z_{\alpha/2} \cdot \sigma, \quad (\bar{x}_1 - \bar{x}_2) + z_{\alpha/2} \cdot \sigma) \qquad (10.26)$$

where $\bar{x}_1$ and $\bar{x}_2$ are the numerical values of $\bar{x}_1$ and $\bar{x}_2$ for the particular samples obtained, and σ is given by (10.24).

Procedure: Confidence Interval for $\mu_1 - \mu_2$ *(Variances Known)*

(1) Decide upon a degree of confidence, $1 - \alpha$.
Find $z_{\alpha/2}$ in standard normal tables.
(2) Obtain a random sample of size n_1 from Population 1.
Calculate $\bar{x}_1$.
(3) Obtain a random sample of size n_2 from Population 2.
Calculate $\bar{x}_2$.
(4) Calculate $\bar{x}_1 - \bar{x}_2$.
Calculate $\sigma = \sqrt{\sigma_1^2/n_1 + \sigma_2^2/n_2}$
(5) Calculate $L = (\bar{x}_1 - \bar{x}_2) - z_{\alpha/2} \cdot \sigma$.
Calculate $U = (\bar{x}_1 - \bar{x}_2) + z_{\alpha/2} \cdot \sigma$.
(6) A $100(1 - \alpha)$ per cent confidence interval for $\mu_1 - \mu_2$ obtained is (L, U).

EXAMPLE: A sample of size 13 from a normal population with variance 100 yielded $\bar{x}_1 = 31\cdot4$. A sample of size 7 from a second normal population with variance 80 yielded $\bar{x}_2 = 38\cdot1$. Find a 95 per cent confidence interval for $\mu_1 - \mu_2$.

(1) $$1 - \alpha = 0\cdot95$$
$$z_{0.025} = 1\cdot96$$

(2) The random sample has already been obtained.
$$\bar{x}_1 = 31\cdot4$$

(3) The random sample has already been obtained.
$$\bar{x}_2 = 38\cdot1$$

(4) $$\bar{x}_1 - \bar{x}_2 = 31\cdot4 - 38\cdot1 = -6\cdot7$$
$$\sigma = \sqrt{\tfrac{100}{13} + \tfrac{80}{7}} = \sqrt{19\cdot12} = 4\cdot37$$

(5) $$L = -6\cdot7 - (1\cdot96)(4\cdot37) = -6\cdot7 - 8\cdot6 = -15\cdot3$$
$$U = -6\cdot7 + (1\cdot96)(4\cdot37) = -6\cdot7 + 8\cdot6 = 1\cdot9$$

(6) The 95 per cent confidence interval is $(-15\cdot3, 1\cdot9)$

(2) *Normal Populations, Common Unknown Variance*

When $\bar{x}_1$ is based on a random sample of size n_1 from a normal population with unknown mean μ_1 and unknown variance σ^2, and $\bar{x}_2$ is based on a random sample of size n_2 from a normal population with unknown mean μ_2 and unknown variance σ^2 (the same variance as the first population), the random variable

$$t = \frac{(\bar{x}_1 - \bar{x}_2) - (\mu_1 - \mu_2)}{\sqrt{\dfrac{s_p^2}{n_1} + \dfrac{s_p^2}{n_2}}} \qquad (10.27)$$

where

$$s_p^2 = \frac{(n_1 - 1)s_1^2 + (n_2 - 1)s_2^2}{n_1 + n_2 - 2} \qquad (10.28)$$

has a *t*-distribution with $n_1 + n_2 - 2$ degrees of freedom. In a particular situation, s_1^2 and s_2^2 are the variances of the two samples; and the numerical value of the random variable, s_p^2, is

$$s_p^2 = \frac{(n_1 - 1)s_1^2 + (n_2 - 1)s_2^2}{n_1 + n_2 - 2} \qquad (10.29)$$

Hereafter, in Case (2), we will denote the random variable

$$\sqrt{\frac{s_p^2}{n_1} + \frac{s_p^2}{n_2}}$$

by **s**. In a particular situation the random variable s_p^2 has the value s_p^2, given in (10.29), and the random variable $\sqrt{\dfrac{s_p^2}{n_1} + \dfrac{s_p^2}{n_2}}$ has the value $\sqrt{\dfrac{s_p^2}{n_1} + \dfrac{s_p^2}{n_2}}$, which will be denoted by s:

$$s = \sqrt{\frac{s_p^2}{n_1} + \frac{s_p^2}{n_2}} \qquad (10.30)$$

Since the random variable given in (10.27) has a *t*-distribution, we can say that

$$P\left\{-t_{\alpha/2} < \frac{(\bar{x}_1 - \bar{x}_2) - (\mu_1 - \mu_2)}{s} < t_{\alpha/2}\right\} = 1 - \alpha \qquad (10.31)$$

where $t_{\alpha/2}$ is the upper $\alpha/2$-point of the *t*-distribution with $n_1 + n_2 - 2$ degrees of freedom. The probability statement in (10.31) can be shown equivalent to the statement

$$P\{(\bar{x}_1 - \bar{x}_2) - st_{\alpha/2} < \mu_1 - \mu_2 < (\bar{x}_1 - \bar{x}_2) + st_{\alpha/2}\} = 1 - \alpha \qquad (10.32)$$

An exact $100(1 - \alpha)$ per cent confidence interval for $\mu_1 - \mu_2$ is given by

$$((\bar{x}_1 - \bar{x}_2) - st_{\alpha/2}, \quad (\bar{x}_1 - \bar{x}_2) + st_{\alpha/2}) \qquad (10.33)$$

where $\bar{x}_1$ and $\bar{x}_2$ are the means of the two samples, s is given by Formula (10.30), and t is the upper $\alpha/2$-point of the *t*-distribution with $n_1 + n_2 - 2$ degrees of freedom.

Procedure: Confidence Interval for $\mu_1 - \mu_2$, *Common Variance Unknown*

(1) Decide upon a degree of confidence, $1 - \alpha$.
    Decide upon the sizes of the samples, n_1 and n_2.
    Find $t_{\alpha/2}(n_1 + n_2 - 2) = t_{\alpha/2}$ in the *t*-table.

(2) Obtain a random sample of size n_1 from Population 1.
   Calculate $\bar{x}_1$.
   Calculate s_1^2.
(3) Obtain a random sample of size n_2 from Population 2.
   Calculate $\bar{x}_2$.
   Calculate s_2^2.
(4) Calculate $\bar{x}_1 - \bar{x}_2$.
   Calculate s_p^2 from (10.29).
   Calculate s from (10.30).
(5) Calculate $L = (\bar{x}_1 - \bar{x}_2) - t_{\alpha/2} \cdot s$.
   Calculate $U = (\bar{x}_1 - \bar{x}_2) + t_{\alpha/2} \cdot s$.
(6) A $100(1 - \alpha)$ per cent confidence interval for $\mu_1 - \mu_2$ is (L, U).

EXAMPLE: A sample from a normal population with unknown variance consists of the observations 34, 25, 43, 37, 45. A sample from a second normal population with the same unknown variance as the first consists of the obervations 20, 31, 23, 35, 41, 29, 39. Find a 95 per cent confidence interval on $\mu_1 - \mu_2$.

(1) $1 - \alpha = 0\cdot95$.
   The sample sizes are $n_1 = 5$ and $n_2 = 7$.

$$t_{0\cdot025}(10) = 2\cdot228$$

(2) The random sample from Population 1 is given.

$$\bar{x}_1 = \frac{184}{5} = 36\cdot8$$

$$s_1^2 = 1/4\left[7024 - \frac{(184)^2}{5}\right] = \tfrac{1}{4}[7024 - 6771\cdot2]$$

$$= \frac{252\cdot8}{4} = 63\cdot2$$

(3) The random sample from Population 2 has already been obtained.

$$\bar{x}_2 = \frac{218}{7} = 31\cdot1$$

$$s_2^2 = 1/6\left[7158 - \frac{(218)^2}{7}\right] = \tfrac{1}{6}[7158 - 6789\cdot1]$$

$$= \frac{368\cdot9}{6} = 61\cdot5$$

(4) $$\bar{x}_1 - \bar{x}_2 = 36\cdot8 - 31\cdot1 = 5\cdot7$$

$$s_p^2 = \frac{(n_1 - 1)s_1^2 + (n_2 - 1)s_2^2}{n_1 + n_2 - 2} = \frac{252\cdot8 + 368\cdot9}{10}$$

$$= 62\cdot17$$

$$s = \sqrt{\frac{62\cdot17}{5} + \frac{62\cdot17}{7}} = \sqrt{12\cdot43 + 8\cdot88}$$

$$= \sqrt{21\cdot31} = 4\cdot62$$

(5) $$L = 5\cdot7 - (2\cdot228)(4\cdot62) = 5\cdot7 - 10\cdot3 = -4\cdot6$$
$$U = 5\cdot7 + (2\cdot228)(4\cdot62) = 5\cdot7 + 10\cdot3 = 16\cdot0$$

(6) The 95 per cent confidence interval for $\mu_1 - \mu_2$ is $(-4\cdot6, 16\cdot0)$.

(3) *Non-normal Populations*

When the sample sizes are large (at least thirty observations each) the populations need not be normal, nor do the variances need to be known.

However, the confidence interval for $\mu_1 - \mu_2$ is only approximate, and not exact as it was in Cases (1) and (2).

When the random variable $\bar{x}_1$ is based on a sample of size n_1 from any population with mean μ_1 and variance σ_1^2, and the random variable $\bar{x}_2$ is based on a sample of size n_2 from any second population with mean μ_2 and variance σ_2^2, then the random variable $\bar{x}_1 - \bar{x}_2$ is approximately normal with mean $\mu_1 - \mu_2$ and variance

$$\sigma^2 = \frac{\sigma_1^2}{n_1} + \frac{\sigma_2^2}{n_2}$$

Therefore, the random variable

$$z = \frac{(\bar{x}_1 - \bar{x}_2) - (\mu_1 - \mu_2)}{\sigma} \qquad (10.34)$$

where

$$\sigma = \sqrt{\frac{\sigma_1^2}{n_1} + \frac{\sigma_2^2}{n_2}} \qquad (10.35)$$

is approximately standard normal.

If the population variances are unknown they may be estimated by the sample variances. For large samples, the sample variances furnish very close estimates of population variances. Therefore, we can say that, for large samples, the random variable

$$z = \frac{(x_1 - x_2) - (\mu_1 - \mu_2)}{s}$$

where

$$s = \sqrt{\frac{s_1^2}{n_1} + \frac{s_2^2}{n_2}}$$

is also approximately standard normal.

Since the random variable z given in (10.34) is standard normal, we know that

$$P\left\{ -z_{\alpha/2} < \frac{(\bar{x}_1 - \bar{x}_2) - (\mu_1 - \mu_2)}{\sigma} < z_{\alpha/2} \right\} = 1 - \alpha$$

This probability statement can be shown equivalent to

$$P\{(\bar{x}_1 - \bar{x}_2) - z_{\alpha/2} \cdot \sigma < \mu_1 - \mu_2 < (\bar{x}_1 - \bar{x}_2) + z_{\alpha/2} \cdot \sigma\} = 1 - \alpha$$

An approximate $100(1 - \alpha)$ per cent confidence interval for $\mu_1 - \mu_2$ when the population variances are known is given by

$$((\bar{x}_1 - \bar{x}_2) - z_{\alpha/2} \cdot \sigma, \ (\bar{x}_1 - \bar{x}_2) + z_{\alpha/2} \cdot \sigma)) \qquad (10.36)$$

where $\bar{x}_1$ and $\bar{x}_2$ are the means of the two samples, and σ is given by (10.35).

When the population variances are unknown, an approximate $100(1 - \alpha)$ per cent confidence interval for $\mu_1 - \mu_2$ is given by

$$((\bar{x}_1 - \bar{x}_2) - z_{\alpha/2} \cdot s, \ (\bar{x}_1 - \bar{x}_2) + z_{\alpha/2} \cdot s) \qquad (10.37)$$

where $\bar{x}_1$ and $\bar{x}_2$ are the means of the two samples, and

$$s = \sqrt{\frac{s_1^2}{n_1} + \frac{s_2^2}{n_2}} \qquad (10.38)$$

where s_1^2 and s_2^2 are the two sample variances.

Procedure: Confidence Interval for $\mu_1 - \mu_2$, *Non-normal Populations*

(1) Decide upon a degree of confidence, $1 - \alpha$.
Find $z_{\alpha/2}$ in standard normal tables.
(2) Obtain a random sample of size n_1 from Population 1.
Calculate $\bar{x}_1$.
If σ_1^2 is unknown, calculate s_1^2.
(3) Obtain a random sample of size n_2 from Population 2.
Calculate $\bar{x}_2$.
If σ_2^2 is unknown, calculate s_2^2.
(4) Calculate $\bar{x}_1 - \bar{x}_2$.
Calculate σ from (10.35) or s from (10.38), whichever is appropriate.
(5) Calculate $L = (\bar{x}_1 - \bar{x}_2) - z_{\alpha/2} \cdot \sigma$.
Calculate $U = (\bar{x}_1 - \bar{x}_2) + z_{\alpha/2} \cdot \sigma$.
If s instead of σ has been calculated in step (4), then s should be used in this step also, in place of σ.
(6) A $100(1 - \alpha)$ per cent confidence interval for $\mu_1 - \mu_2$ is (L, U).

EXAMPLE: A sample of size 30 was obtained from a first population, and it was found that $\bar{x}_1 = 305$ and $s_1^2 = 1548$. Another sample, of size 43, was drawn from a second population, and it was found that $\bar{x}_2 = 289$ and $s_2^2 = 1654$. Find the 99 per cent confidence interval for $\mu_1 - \mu_2$ that is determined by these data.

(1) We were given $1 - \alpha = 0.99$; $z_{0.005} = 2.58$.
(2) The random sample from the first population has already been obtained.

$$\bar{x}_1 = 305$$
$$s_1^2 = 1548$$

(3) The random sample from the second population has already been obtained:

$$\bar{x}_2 = 289$$
$$s_2^2 = 1654$$

(4) $\quad \bar{x}_1 - \bar{x}_2 = 305 - 289 = 16$

$$s = \sqrt{\tfrac{1548}{30} + \tfrac{1654}{43}} = \sqrt{51.60 + 38.47}$$
$$= \sqrt{90.07} = 9.49$$

(5) $\quad L = 16 - (2.58)(9.49) = 16 - 24.5 = -8.5$
$\quad U = 16 + (2.58)(9.49) = 16 + 24.5 = 40.5$

(6) The 99 per cent confidence interval determined by the given data is $(-8.5, 40.5)$.

CONFIDENCE INTERVAL FOR $\pi_1 - \pi_2$

The true proportion of successes in one binomial population is π_1, and the true proportion in a second binomial population is π_2. If the random variable p_1 is based on a random sample of size n_1 from the first population, and the random variable p_2 is based on a random sample of size n_2 from the second population, then the random variable $p_1 - p_2$ has mean $\pi_1 - \pi_2$ and variance

$$\frac{\pi_1(1 - \pi_1)}{n_1} + \frac{\pi_2(1 - \pi_2)}{n_2}$$

The standardized random variable

$$z = \frac{(p_1 - p_2) - (\pi_1 - \pi_2)}{\sqrt{\dfrac{\pi_1(1 - \pi_1)}{n_1} + \dfrac{\pi_2(1 - \pi_2)}{n_2}}}$$

is approximately standard normal. Therefore, we can make the approximate probability statement,

$$P\left\{ -z_{\alpha/2} < \frac{(p_1 - p_2) - (\pi_1 - \pi_2)}{\sqrt{\dfrac{\pi_1(1 - \pi_1)}{n_1} + \dfrac{\pi_2(1 - \pi_2)}{n_2}}} < z_{\alpha/2} \right\} \cong 1 - \alpha \quad (10.39)$$

After the usual manipulation of the continued inequality, (10.39) becomes

$$P\{(p_1 - p_2) - z_{\alpha/2} \cdot \sigma < \pi_1 - \pi_2 < (p_1 - p_2) + z_{\alpha/2} \cdot \sigma\} \cong 1 - \alpha \quad (10.40)$$

where

$$\sigma = \sqrt{\frac{\pi_1(1 - \pi_1)}{n_1} + \frac{\pi_2(1 - \pi_2)}{n_2}} \quad (10.41)$$

A $100(1 - \alpha)$ per cent confidence interval for $\pi_1 - \pi_2$ is given by

$$((\hat{p}_1 - \hat{p}_2) - z_{\alpha/2} \cdot \sigma, \quad (\hat{p}_1 - \hat{p}_2) + z_{\alpha/2} \cdot \sigma) \quad (10.42)$$

where the number $\hat{p}_1$ is the proportion observed in the first sample, the number $\hat{p}_2$ is the proportion observed in the second sample, and σ is given in (10.41). The trouble with this interval, of course, is that it cannot be determined, because, in order to find the value of σ, the values of the unknown π_1 and π_2 are needed.

In order to obtain a confidence interval for $\pi_1 - \pi_2$, we must either know or estimate the quantity

$$\frac{\pi_1(1 - \pi_1)}{n_1} + \frac{\pi_2(1 - \pi_2)}{n_2} \quad (10.43)$$

We can estimate the population proportion π_1 by the sample proportion $\hat{p}_1$; and we can estimate the population proportion π_2 by the sample proportion $\hat{p}_2$. Therefore, our estimate of the true variance of $p_1 - p_2$, the quantity in (10.43), is

$$\frac{\hat{p}_1(1 - \hat{p}_1)}{n_1} + \frac{\hat{p}_2(1 - \hat{p}_2)}{n_2} \quad (10.44)$$

we will denote the square root of this quantity by s.

Thus, an approximate $100(1 - \alpha)$ per cent confidence interval for $\pi_1 - \pi_2$ is given by

$$((\hat{p}_1 - \hat{p}_2) - z_{\alpha/2} \cdot s, \quad (\hat{p}_1 - \hat{p}_2) + z_{\alpha/2} \cdot s) \quad (10.45)$$

where $\hat{p}_1$ and $\hat{p}_2$ are the observed proportions for the two particular samples obtained, and s is the square root of the quantity in (10.44).

Procedure: Confidence Interval for $\pi_1 - \pi_2$

    (1) Decide upon a degree of confidence, $1 - \alpha$.
        Find $z_{\alpha/2}$ in standard normal tables.

(2) Obtain a random sample of size n_1 from Population 1.
   Calculate $\hat{p}_1$ (three-place accuracy is recommended).

(3) Obtain a random sample of size n_2 from Population 2.
   Calculate $\hat{p}_2$ (three-place accuracy is recommended).

(4) Calculate $\hat{p}_1 - \hat{p}_2$.

   Calculate $s = \sqrt{\dfrac{\hat{p}_1(1 - \hat{p}_1)}{n_1} + \dfrac{\hat{p}_2(1 - \hat{p}_2)}{n_2}}$

(5) Calculate $L = (\hat{p}_1 - \hat{p}_2) - z_{\alpha/2} \cdot s$
   $U = (\hat{p}_1 - \hat{p}_2) + z_{\alpha/2} \cdot s.$

(6) A $100(1 - \alpha)$ per cent confidence interval for $\pi_1 - \pi_2$ is (L, U).

EXAMPLE: Tom and Joe like to throw darts. Tom throws 100 times and hits the target 54 times; Joe throws 100 times and hits the target 49 times. Find a 95 per cent confidence interval for $\pi_1 - \pi_2$, where π_1 represents the true proportion of hits in Tom's tosses, and π_2 represents the true proportion of hits in Joe's tosses.

(1) The degree of confidence is 0.95; $z_{0.025} = 1.96$.

(2) $$\hat{p}_1 = \frac{54}{100} = 0.540$$

(3) $$\hat{p}_2 = \frac{49}{100} = 0.490$$

(4) $$\hat{p}_1 - \hat{p}_2 = 0.540 - 0.490 = 0.050$$

$$s = \sqrt{\frac{(0.54)(0.46)}{100} + \frac{(0.49)(0.51)}{100}}$$

$$= \sqrt{\frac{0.2484}{100} + \frac{0.2499}{100}} = \sqrt{\frac{0.4983}{100}}$$

$$= \sqrt{0.004983} = 0.0706$$

(5) $\quad L = 0.050 - (1.96)(0.0706) = 0.050 - 0.138 = -0.088$
   $\quad U = 0.050 + (1.96)(0.0706) = 0.050 + 0.138 = 0.188$

(6) A 95 per cent confidence interval for $\pi_1 - \pi_2$ is $(-0.088, 0.188)$.

CONFIDENCE INTERVAL FOR ρ

From Chapter IX we know that the random variable known as the Fisher-z, z_f, is approximately normal with mean μ_f and variance $\dfrac{1}{n-3}$. The relationship between z_f and r, the sample correlation coefficient, is given in Formula (9.3); the Fisher-z table is used to find Fisher-z values that correspond to particular r-values, and vice versa.

In order to find a confidence statement for ρ, the population correlation coefficient, we must first find a confidence interval for μ_f, and then transform this interval into an interval for ρ. We can begin, since the Fisher-z is approximately normal with the mean and variance given above, with the statement

$$P\left\{-z_{\alpha/2} < \frac{z_f - \mu_f}{\dfrac{1}{\sqrt{n-3}}} < z_{\alpha/2}\right\} \cong 1 - \alpha \qquad (10.46)$$

(Do not confuse the z's: $z_{\alpha/2}$ is the upper $\alpha/2$-point of the standard normal distribution; z_f is the Fisher-z random variable.) After the usual manipulation of the continued inequality within braces, the statement in (10.46) becomes

$$P\left\{z_f - \frac{z_{\alpha/2}}{\sqrt{n-3}} < \mu_f < z_f + \frac{z_{\alpha/2}}{\sqrt{n-3}}\right\} \cong 1 - \alpha \qquad (10.47)$$

For a particular sample, we have a particular value of $\mathbf{r}$ and, therefore, a particular value of z_f; a $100(1 - \alpha)$ per cent confidence interval for μ_f is given by

$$\left(z_f - \frac{z_{\alpha/2}}{\sqrt{n-3}}, z_f + \frac{z_{\alpha/2}}{\sqrt{n-3}}\right) \qquad (10.48)$$

We are interested in a confidence interval for μ_f, however, only as a means by which a confidence interval for ρ may be obtained. The number $z_f - \frac{z_{\alpha/2}}{\sqrt{n-3}}$, considered as a Fisher-z value, determines a value of the correlation coefficient that will be denoted ρ_L. Also, the number $z_f + \frac{z_{\alpha/2}}{n-3}$, considered as a Fisher-z value, determines a value of the correlation coefficient that will be denoted by ρ_U. Whenever μ_f is contained in the interval given in (10.48), ρ is contained in the interval (ρ_L, ρ_U).

For example, suppose that a 90 per cent confidence interval for μ_f is (0·55, 1·18). Referring to the Fisher-z table, we see that the value of r that corresponds to 0·55 is 0·50. Also, the value of r that corresponds to 1·18 is 0·83. Therefore, if (0·55, 1·18) is a 90 per cent confidence interval for μ_f, then (0·50, 0·83) is a 90 per cent confidence interval for ρ.

The procedure for obtaining a confidence interval for ρ is given in more detail below.

Procedure: Confidence Interval for ρ

   (1) Decide upon a degree of confidence, $1 - \alpha$.
       Find $z_{\alpha/2}$ in standard normal tables.
   (2) Obtain a random sample of size n.
       Calculate r, the sample correlation coefficient.
       Find z_f, the Fisher-z value, that corresponds to the value of r. (Use the Fisher-z table.)
   (3) Calculate $L = z_f - \frac{z_{\alpha/2}}{\sqrt{n-3}}$
       Find ρ_L, the value that corresponds to a Fisher-z value equal to L. (Use the Fisher-z table.)
   (4) Calculate $U = z_f + \frac{z_{\alpha/2}}{\sqrt{n-3}}$
       Find ρ_U, the value that corresponds to a Fisher-z value equal to U. (Use the Fisher-z table.)
   (5) A $100(1 - \alpha)$ per cent confidence interval for ρ is (ρ_L, ρ_U).

EXAMPLE: A sample of size 28 yielded $r = 0.63$. Find a 95 per cent confidence interval for ρ.

(1) The degree of confidence is 0.95.

$$z_{0.025} = 1.96$$

(2) The random sample has already been obtained. We are given $r = 0.63$. The Fisher-z value that corresponds to $r = 0.63$ is 0.74.

(3) $\qquad L = 0.74 - \dfrac{1.96}{\sqrt{25}} = 0.74 - \dfrac{1.96}{5} = 0.74 - 0.39 = 0.35$

$\qquad \rho_L = 0.34$

(4) $\qquad\qquad U = 0.74 + \dfrac{1.96}{\sqrt{25}} = 0.74 + 0.39 = 1.13$

$\qquad\qquad \rho_U = 0.81$

(5) A 95 per cent confidence interval for ρ is $(0.34, 0.81)$.

EXERCISES

1. Given that a sample of size 16 from a normal distribution yielded $\bar{x} = 25$. The variance is known to be 64. Find a 90 per cent confidence interval for μ; a 95 per cent interval for μ; a 99 per cent interval for μ.

2. A sample of size 50 from a population yielded the sample values $\bar{x} = 190$ and $s^2 = 800$. Find a 95 per cent confidence interval for μ.

3. A sample of size 9 from a normal population gave $\bar{x} = 15.8$ and $s^2 = 10.3$. Find a 99 per cent interval for μ.

4. A man flips a coin that he believes is biased 50 times, and observes 35 heads. Find an approximate 90 per cent confidence interval for π, the true proportion of heads.

5. If a nationwide sample of size 1000 contains 550 people that support candidate for election Mr. A, what is a 90 per cent confidence interval for π, the true proportion of the nation's voters that support Mr. A? A 95 per cent interval? A 99 per cent interval?

6. The results of two samples from two normal populations with equal variances are the following: For Sample 1, $n_1 = 11$, $\bar{x}_1 = 53$, $s_1^2 = 110$. For Sample 2, $n_2 = 7$, $\bar{x}_2 = 49$, and $s_2^2 = 130$. Find a 90 per cent confidence interval for $\mu_1 - \mu_2$.

7. A random sample of size 10 from a normal population with variance 50 gave $\bar{x}_1 = 43.2$. A second random sample of size 18 from a normal population with variance 72 gave $\bar{x}_2 = 48.7$. Find a 99 per cent confidence interval for $\mu_1 - \mu_2$.

8. A random sample of size 100 yielded the sample values $\bar{x}_1 = 509$, $s_1^2 = 950$. A random sample of size 100 from another population yielded $\bar{x}_2 = 447$ and $s_2^2 = 875$. Find a 95 per cent interval for $\mu_1 - \mu_2$.

9. Find a 95 per cent confidence interval for $\pi_1 - \pi_2$, if a sample of size 200 yielded $\hat{p}_1 = 0.40$, and a sample of size 100 yielded $\hat{p}_2 = 0.30$.

10. Find a 90 per cent confidence interval for $\pi_1 - \pi_2$, if a sample of size 20 yielded $\hat{p}_1 = 0.70$, and a sample of size 30 yielded $\hat{p}_2 = 0.65$.

11. Work the preceding problem if the sample sizes were 200 and 300, instead of 20 and 30.

12. A sample of size 12 gave $r = 0.35$. Find a 99 per cent interval for ρ.

13. A sample of size 19 gave $r = -0.75$. Find a 90 per cent interval for ρ.

14. A sample of size 12 gave $r = 0.60$. Find a 95 per cent interval for ρ. Find a 95 per cent interval for ρ if the sample size was 103. Find a 95 per cent interval for ρ if the sample size was 403.

CHAPTER XI

NON-PARAMETRIC STATISTICS

INTRODUCTION

Most of the tests presented in Chapters VII and VIII could not be made unless we assumed that we were sampling from a normal distribution or a binomial distribution. All of the tests were about parameters—population means, variances, and proportions.

This chapter is concerned with tests known as **non-parametric tests**. The first three sections discuss some additional uses of the χ^2 distribution. The type of hypothesis tested here is quite different from the hypotheses tested previously: we learn how to test the independence of two criteria of classification and how to test whether the population from which we are sampling is normal. The rank correlation coefficient, which is used to test whether two variables x and y are independent, is discussed.

The remainder of the chapter deals with non-parametric tests about the median of a population, or the medians of two populations. Two important advantages of many non-parametric tests are that they do not require many assumptions and that they are relatively quick and easy to apply, requiring few computations. Therefore, in testing situations in which the normality assumption cannot be satisfied and the sample size is not large enough to warrant appeal to the Central Limit Theorem, one of the non-parametric tests about the median will do very nicely.

THE CHI-SQUARED DISTRIBUTION

There are many experiments in which each member of the sample is observed and classified in one of several different categories, known as **cells**. For instance, if a hybrid male rat (the chromosomes contain a dominant and a recissive gene for some particular characteristic) is mated to a hybrid female rat each offspring can be classified into one of three categories: (1) the offspring received a dominant gene from each parent; (2) the offspring received a dominant gene from one parent and a recessive gene from the other parent; and (3) the offspring received a recessive gene from each parent.

When a pair of dice is rolled a thousand times the size of the sample is 1000 and the members are the individual rolls. The experiment of throwing a pair of dice can result in any one of eleven outcomes—totals of 2 through 12. Each roll can be classified into one of the eleven categories, depending upon the total obtained on that roll.

There are four blood types—A, O, B, and AB. Suppose that a random sample of 100 people is taken. Each person can be classified according to his blood type into one of four categories: one category corresponds to Type A, one to Type B, one to Type O, and one to Type AB.

Frequently it is desired to test whether the theoretical proportions in each cell are equal to specified numerical values. (The sum of the theoretical proportions, like the sum of theoretical probabilities, must be 1.) For example, suppose that a die is tossed 120 times and the result recorded for each toss. The 120 tosses might yield the results

Number of spots	1	2	3	4	5	6
Observed frequency	16	19	27	17	23	18

If the die is honest the theoretical proportion in each of the six cells is $\frac{1}{6}$ (equivalent to saying that the probability of throwing any particular number of spots is $\frac{1}{6}$); to find the **theoretical frequency** (or number) in each cell we multiply the total number of trials (the sample size) by the theoretical proportions. In 120 tosses the theoretical number of ones is 20; the theoretical number of twos is 20; and so on. The theoretical number is usually referred to as the **expected number**. When we say that the expected number of ones is 20 we do not mean that we would be amazed if we threw a die 120 times and did not obtain exactly 20 ones. We mean that it is reasonable to anticipate obtaining *about* 20 ones, and we would be surprised, perhaps, if we get a number of ones very different from 20. If the experiment of throwing an honest die 120 times is performed a great many times the average number of ones obtained will be very nearly 20; and the more times the experiment is performed, the closer to 20 the average number of ones will be. Similar remarks, of course, could be made about the number of twos, or threes, or sixes. So, although we *do* 'expect' to obtain a number of ones that is *about* 20, the term 'expected number' simply means 'theoretical number', and does not refer to a number that we are certain to observe.

The observed and expected numbers (frequencies) are usually displayed as in Table 11.1.

Table 11.1

No. of spots	1	2	3	4	5	6
Observed No.	16	19	27	17	23	18
Expected No.	20	20	20	20	20	20

The hypothesis that the die is honest is equivalent to the hypothesis that the expected proportion in each cell is $\frac{1}{6}$, which is equivalent to the hypothesis that each expected number is 20, where 120 is the total sample size. If the numbers that we observe are very different from those that we think we should observe when the die is honest we will decide that the die is *not* honest. The question is: how different is 'very different'? As usual, we need a numerical quantity to help us in making our decisions.

We see (in Table 11.1) that the observed number of ones was 16, while the expected number was 20. One way to measure the difference between observed and expected numbers is simply to subtract the expected number from the observed number; this would yield -4 in this case. If we did the same for each of the other cells and added the 6 resulting differences this could be interpreted as a measure of how much the observed frequencies differ from expected frequencies.

However, the sum of the deviations is always zero, no matter how great individual deviations might be; so the sum of the deviations is not a satisfactory measure. For instance, the differences would sum to zero if the observed frequencies in 120 tosses were those shown in Table 11.2, yet obviously the observed frequencies differ from the expected.

Table 11.2

Observed No.	10	30	10	30	10	30
Expected No.	20	20	20	20	20	20

A more satisfactory procedure would be to square each difference and then add the resulting squares of differences. This would always yield a positive number, and this number would be larger when the deviations of observed from expected frequencies are large than it would be when they are small. For the data in Table 11.1 the sum of the squared differences is

$$(-4)^2 + (-1)^2 + 7^2 + (-3)^2 + 3^2 + (-2)^2$$
$$= 16 + 1 + 49 + 9 + 9 + 4$$
$$= 88$$

It would seem, then, that the sum of the squared differences would be a suitable numerical measure of the deviation of observed from expected values. There is an objection, however, to the use of the squared differences.

Consider the two sets of data displayed in Tables 11.3 and 11.4.

Table 11.3

Observed	13	12	7
Expected	6	15	11

Table 11.4

Observed	457	297	496
Expected	450	300	500

The reader may verify that the sum of squares of differences for each table is 74. However, it seems reasonable to think that a discrepancy of 7 is more significant when the expected number is 6 and the observed is 13, than when the expected number is 250 and the observed is 257, even though the squared difference is 49 in both cases. In the first case the observed is *more than twice* the expected; in the second the observed number is only about 3 per cent greater than the expected. What is important is not the size of the squared difference, namely 49, but the *size of the squared difference relative to the expected frequency*, which is found by dividing the squared difference by the expected frequency. This relative squared difference for the first cell of Table 11.3 is $\frac{49}{6} = 8.17$; for the first cell of Table 11.4 it is $\frac{49}{450} = 0.11$.

To get a measure of the discrepancy between observed and expected frequencies, we will add the relative squared differences for each cell of the table. Thus, the measure of discrepancy for the data in Table 11.3 is

$$\frac{49}{6} + \frac{9}{15} + \frac{16}{11} = 8.17 + 0.60 + 1.45 = 10.22$$

and the measure for the data in Table 11.4 is

$$\frac{49}{450} + \frac{9}{300} + \frac{16}{300} = 0.11 + 0.03 + 0.05 = 0.19$$

Now we have two measures of discrepancy values, 10·22 and 0·19, that are very different, whereas before both measures equalled 74.

The sum of the relative squared differences is called chi-squared and written χ^2. Denoting the observed frequency of the ith class by O_i and the expected frequency of the ith class by E_i, the statistic χ^2 is defined by the formula

$$\chi^2 = \sum \frac{(O_i - E_i)^2}{E_i} \qquad (11.1)$$

and is a measure of the discrepancy between observed and expected frequencies. The larger χ^2 is, the less agreement there is between observed and expected frequencies. The smaller χ^2 is, the more closely the observed and expected frequencies agree.

EXAMPLE: Using the data in Table 11.1, calculate χ^2.

We need $\sum \dfrac{(O_i - E_i)^2}{E_i}$. The necessary computations are shown in Table 11.5.

Table 11.5

O_i	E_i	$O_i - E_i$	$(O_i - E_i)^2$	$\dfrac{(O_i - E_i)^2}{E_i}$
16	20	−4	16	0·80
19	20	−1	1	0·05
27	20	7	49	2·45
17	20	−3	9	0·45
23	20	3	9	0·45
18	20	−2	4	0·20
$\sum$: 120	120			4·40

The value of χ^2 is the total of the entries in the final column, and for this example is 4·40. It is advisable to sum both the observed frequency column and the expected frequency column, to make sure that the sums agree. (The sum of the expected frequencies might be slightly different from the sample size because of rounding error.)

If we expand (11.1) we obtain an alternative formula for χ^2.

$$\chi^2 = \sum \left\{ \frac{[O_i^2 - 2O_iE_i + E_i^2]}{E_i} \right\}$$

$$= \sum \frac{O_i^2}{E_i} - 2\sum O_i + \sum E_i$$

$$= \sum \frac{O_i^2}{E_i} - 2n + n$$

$$\chi^2 = \left(\sum \frac{O_i^2}{E_i} \right) - n \qquad (11.2)$$

where O_i and E_i are defined as before, and n is the total sample size. The computations necessary to find $\sum \dfrac{O_i^2}{E_i}$ for the data in Table 11.1 are shown in Table 11.6.

Table 11.6

O_i	O_i^2	E_i	$\dfrac{O_i^2}{E_i}$
16	256	20	12·80
19	361	20	18·05
27	729	20	36·45
17	289	20	14·45
23	529	20	26·45
18	324	20	16·20
$\sum$: 120		120	124·40

Substituting the value of $\sum \dfrac{O_i^2}{E_i}$ found in Table 11.6 and the value of n into Formula (11.2), we have

$$\chi^2 = 124\cdot40 - 120 = 4\cdot40$$

which agrees with the value of χ^2 that we obtained using Formula (11.1).

There are various outcomes of the experiment of rolling the die 120 times—they range from obtaining 120 ones to obtaining 20 of each number to obtaining 120 sixes. If the χ^2-value as defined in Formula (11.1) is calculated for each of these possible combinations of observed frequencies we then have a great number of χ^2-values. They would range in value from 0, which indicates perfect agreement of observed and expected frequencies, to 720, which is the largest possible value (in this particular example) and indicates the poorest possible agreement. (Note that χ^2 can never be negative.) If the χ^2-values are classified and the histogram of the resulting frequency distribution drawn it might look like the histogram in Fig. 11.1.

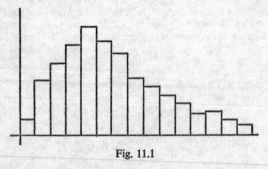

Fig. 11.1

Suppose that 5 per cent of the area in the histogram lies to the right of the value 11·07. Then we could say that the hypothesis that the die is balanced will be rejected if the sample results yield a χ^2-value greater than 11·07. A χ^2-value greater than 11·07 could be obtained purely by chance, even though the die is perfectly balanced; however, we will choose to believe that obtaining such a value is sufficiently unlikely that we will reject the null hypothesis that the die is balanced. The upper 5 per cent point can be calculated straightforwardly by mathematical methods, but it is extremely tedious unless the sample size is very small. Fortunately, there is a distribution, called the **chi-squared distribution**, whose graph closely resembles the histogram in Fig. 11.1. One parameter, the number of degrees of freedom, is sufficient to characterize the distribution. Tables of the distribution are widely available, and one is included in the Appendix of this book.

The χ^2 table is entered with the appropriate number of degrees of freedom and the desired α-level. The α-values are 0·995, 0·99, 0·975, 0·95, 0·05, 0·025, 0·01, and 0·005, and are the column headings for the eight columns of the table. The degrees of freedom are consecutive from 1 to 30. The symbol $\chi^2_\alpha(n)$ denotes the χ^2-value to the right of which lies $100(1 - \alpha)$ per cent of the area under the graph of the χ^2 distribution with n degrees of freedom (see Fig. 11.2).

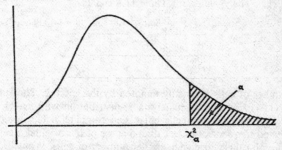

Fig. 11.2

To find a particular $\chi^2_\alpha(n)$, read down the column headed by the particular α-value until the row corresponding to the particular value of n is reached; there read the value of $\chi^2_\alpha(n)$ that is desired.

EXAMPLE: Find $x^2{}_{0.05}(17)$.
In the column headed by 0·05 we read down until we are in the row opposite 17. At that point we find the number 27·587. Therefore,

$$\chi^2{}_{0.05}(17) = 27{\cdot}587$$

The reader may verify for practice the following χ^2 values:

$$\chi^2{}_{0.995}(4) = 0{\cdot}207$$
$$\chi^2{}_{0.01}(23) = 41{\cdot}638$$
$$\chi^2{}_{0.95}(9) = 3{\cdot}325$$

A hypothesis that specifies the theoretical proportions that are contained in each of several categories is known as a **multinomial hypothesis**. In order to obtain a valid approximation to the distribution of the quantity

$$\chi^2 = \sum \left\{ \frac{(O_i - E_i)^2}{E_i} \right\}$$

by using the χ^2-distribution, the expected frequency in each category (cell) must be at least 5. If one or more cells have expected frequencies of less than

5 they must be combined with others until new cells are formed whose expected frequencies are all at least 5. Usually adjacent cells are combined, although the cells may be combined in any manner that does not deliberately attempt to influence the differences between observed and expected frequencies, and thereby influence the value of χ^2.

In order to illustrate these remarks, we will consider Table 11.7.

Table 11.7

O_i:	4	0	7	5	7	1
E_i:	2	2	3	7	9	1

Cells 1, 2, 3, and 6 have expected frequencies that are less than 5. In order to form a new cell with an expected frequency of at least 5, cells 1, 2, and 3 may be combined, and cell 6 may be combined with cell 5. The observed frequencies are also combined in the same manner. Thus, we replace Table 11.7 by Table 11.8. The χ^2-value for Table 11.8 is 3·26.

Table 11.8

O_i:	11	5	8
E_i:	7	7	10

The number of cells is usually denoted by the letter k. The statistic χ^2 in Formula (11.1) has (approximately) a χ^2-distribution with $k - 1$ degrees of freedom (assuming that no cells have been combined). If some of the cells have expected frequencies of less than 5 some of the original k cells must be combined with others to give a smaller number of cells. We will denote this smaller number by k' ('kay-prime'). When it is necessary to combine cells this is done before χ^2 is calculated, and χ^2 then has (approximately) a χ^2-distribution with $k' - 1$ degrees of freedom.

Just as we denoted the hypothetical mean of a population by μ_0, the hypothetical variance by σ_0^2, and the hypothetical proportion of successes in a binomial experiment by π_0, we here denote the hypothetical proportion in the first cell by π_{10}, in the second cell by π_{20}, ... in the kth cell by π_{k0}. The true values of the theoretical proportions are denoted $\pi_1, \pi_2, \ldots \pi_k$.

The hypothesis that is tested in a multinomial hypothesis is that $\pi_1 = \pi_{10}$, $\pi_2 = \pi_{20}, \ldots \pi_k = \pi_{k0}$—that the true value of the theoretical proportion in the first cell equals the numerical value π_{10}, and so on. In the experiment of tossing a die 120 times, the null hypothesis tested would be that $\pi_1 = \frac{1}{6}$, $\pi_2 = \frac{1}{6}, \ldots \pi_6 = \frac{1}{6}$. The procedure for testing a multinomial hypothesis is shown below.

Procedure for Testing a Multinomial Hypothesis

(1) Formulate the null and alternative hypotheses:

H_0: $\pi_1 = \pi_{10}, \pi_2 = \pi_{20}, \ldots \pi_k = \pi_{k0}$.
H_1: At least two of the theoretical proportions are significantly different from the values hypothesized by H_0.

(2) Choose an α-level.

(3) Compute the expected frequencies by multiplying each hypothetical proportion by the sample size, n. The expected frequencies are

$$E_1 = n\pi_{10}, \; E_2 = n\pi_{20}, \; \ldots \; E_k = n\pi_{k0}$$

(4) If any of the expected frequencies found in step (3) are less than 5, combine those cells with each other or with other cells so that each expected frequency is at least 5.

Then calculate $$\chi^2 = \sum \left\{ \frac{(O_i - E_i)^2}{E_i} \right\}$$

(5) The original number of cells was k. If no cells were combined, the number of degrees of freedom for χ^2 is $k - 1$. If at least one cell was combined with others, there are k' cells, and the number of degrees of freedom is $k' - 1$. The number of degrees of freedom is *one less than the number of cells after any necessary combinations of cells have been performed.*

Compare the χ^2-value found in (4) with the upper α-point of the χ^2-distribution with the appropriate degrees of freedom, χ_α^2. Reject H_0 if $\chi^2 \geqslant \chi_\alpha^2$.

EXAMPLE: Experience in a certain apple-growing region indicates that about 20 per cent of the total apple crop is classified in the best grade, about 40 per cent in the next highest rating, about 30 per cent in the next highest rating, and the remaining 10 per cent in the fourth and lowest rating. A random sample of 50 trees in a certain orchard yielded 600 bushels of apples, classified as follows: 150 bushels of the highest-quality apples. 290 bushels of the next highest, 140 bushels of the next highest, and 20 bushels of the lowest-quality apples. Can we conclude that the apples in this orchard are typical?

(1) We will represent the proportion of apples in the four categories by π_1, π_2, π_3, and π_4. The null hypothesis is

$$H_0: \pi_1 = 0{\cdot}20, \; \pi_2 = 0{\cdot}40, \; \pi_3 = 0{\cdot}30, \; \pi_4 = 0{\cdot}10$$

and the alternative hypothesis is

H_1: At least two of the proportions are different from the values specified in H_0.

(2) Let $\alpha = 0{\cdot}01$.

(3)
$$n = 600$$
$$E_1 = 600(0{\cdot}20) = 120$$
$$E_2 = 600(0{\cdot}40) = 240$$
$$E_3 = 600(0{\cdot}30) = 180$$
$$E_4 = 600(0{\cdot}10) = 60$$

(4) The calculation of χ^2 using Formula (11.1) is shown in Table 11.9.

Table 11.9

O_i	E_i	$O_i - E_i$	$(O_i - E_i)^2$	$\dfrac{(O_i - E_i)^2}{E_i}$
150	120	30	900	7·50
290	240	50	2500	10·42
140	180	−40	1600	8·89
20	60	−40	1600	26·67

$$53{\cdot}48 = \sum \left\{ \frac{(O_i - E_i)^2}{E_i} \right\}$$

(5) The number of classes is 4.
The number of degrees of freedom is 3.

$$\chi^2_{0.01}(3) = 11 \cdot 345$$
$$53 \cdot 48 > 11 \cdot 345$$

Reject H_o and accept H_1.

CONTINGENCY TABLES

If each member of a sample can be classified according to two criteria of classification the results of an experiment can be displayed in a table known as a **contingency table**.

For example, suppose that in a certain study each person is classified by the colour of his hair and by the colour of his eyes. Each person, then, will be classified according to two criteria—hair colour and eye colour. We will call 'eye colour' criterion 1, and label the rows of our table with the different 'levels' of this first criterion. We will call 'hair colour' criterion 2, and label the columns of our table with the different 'levels' of the second criterion. Each person will be placed into one of the four rows and into one of the four columns. The four rows and the four columns form sixteen cells in the table itself. Each person is classified into one of the sixteen cells. Suppose that 600 persons were examined, and the results were those shown in Table 11.10.

Table 11.10

		Criterion 2 (Hair Colour)				
		Blond	Brown	Black	Red	Totals
	Blue	60	40	60	40	200
Criterion 1	Grey	20	50	20	10	100
(Eye Colour)	Hazel	10	50	10	30	100
	Brown	10	160	10	20	200
	Totals	100	300	100	100	600

The hypothesis to be tested here is that the two criteria of classification are independent—that eye colour is unrelated to hair colour, and vice versa. The alternative hypothesis is that the two criteria are *not* independent.

As in the previous section, in order to test the null hypothesis we must compare the frequencies that were actually observed with the frequencies that we think should have been observed if the null hypothesis were true. This comparison of observed and theoretical frequencies is once again effected by means of the quantity χ^2, which may be described as the sum of the relative squared differences between observed and expected frequencies.

Let R_1 denote the total of the observed frequencies in the ith row: thus, R_1 is the total observed in the first row; R_2 is the total observed in the second row; and so on. Denote the total of observed frequencies in the jth column by C_j: C_1 is the total observed in the first column; C_2 is the total observed in the second column; and so on.

Assuming that the null hypothesis of independence of the two criteria is true, the expected frequencies are found as follows: to find the expected frequency of a particular cell, multiply the total of the row in which that cell appears by the total of the column in which that cell appears, and then divide the resulting product by the grand total, n. (In Table 11.10, $n = 600$.)

$$E_{ij} = \frac{R_i \cdot C_j}{n} \qquad (11.3)$$

The cell that is in the ith row and the jth column of the contingency table is called the (i, j)th cell. The observed frequency in the (i, j)th cell will be denoted O_{ij}, and the expected frequency will be denoted by E_{ij}. Thus the quantity χ^2, the sum of the relative squared deviations, is given by the formula

$$\chi^2 = \sum \left\{ \frac{(O_{ij} - E_{ij})^2}{E_{ij}} \right\}$$

where the summation is to be taken over all the cells in the contingency table.

The observed number of persons with blue eyes and blond hair is 60. To find the expected number we multiply 200 (the total in the first row) by 100 (the total in the first column), and divide the resulting product by 600. We have

$$E_{11} = \frac{(200)(100)}{600} = 33 \cdot 3$$

Theoretical frequencies in other cells are found similarly. For instance, the theoretical frequency in the $(3, 2)$th cell (the cell in the third row and second column, corresponding to hazel eyes and brown hair) is

$$E_{32} = \frac{(100)(300)}{600} = 50$$

A convenient way to display the observed and theoretical frequencies is to write both numbers in each cell, distinguishing the theoretical frequencies by writing them with ink of a different colour, circling them, enclosing them in parentheses, or by some similar device. We will enclose the expected values in parentheses. The contingency table displaying the observed and theoretical frequencies is shown in Table 11.11. (The expected values are calculated accurate to tenths so that the round-off error will not be great.)

A convenient check on all computations involved in finding the expected frequencies is made by adding the expected frequencies in each row and

Table 11.11

		Criterion 2 (Hair Colour)				
		Blond	Brown	Black	Red	Totals
	Blue	60	40	60	40	200
		(33·3)	(100·0)	(33·3)	(33·3)	(199·9)
	Grey	20	50	20	10	100
		(16·7)	(50·0)	(16·7)	(16·7)	(100·1)
Criterion 1	Hazel	10	50	10	30	100
(Eye Colour)		(16·7)	(50·0)	(16·7)	(16·7)	(100·1)
	Brown	10	160	10	20	200
		(33·3)	(100·0)	(33·3)	(33·3)	(199·9)
	Totals	100	300	100	100	600
		(100·0)	(300·0)	(100·0)	(100·0)	(600·0)

column. The row totals of expected frequencies should match closely the row totals for observed frequencies. Also, the column totals of expected frequencies should be very near to the column totals for observed frequencies. Any small difference between observed frequency totals and the corresponding expected frequency totals will probably be due to rounding error. For instance, in row 1 of Table 11.11 we see that the observed total is 200 and the expected total is 199·9; the 0·1 difference is due to rounding error.

Upon examination of the totals for the remaining three rows, and the totals for the four columns, we see that the totals of expected frequencies agree closely with the totals of observed frequencies.

In general, if there are r categories of the first criterion of classification and c categories of the second the contingency table will consist of r rows and c columns. The r row totals will be denoted $R_1, R_2, \ldots R_r$. The c column totals will be denoted $C_1, C_2, \ldots C_c$. The number of degrees of freedom for the χ^2 statistic, used to test the null hypothesis that the criteria of classification are independent, is $(r-1)(c-1)$. The same restriction must be placed on the expected frequencies as before—the expected frequency in each cell must be at least 5. If any cells have expected frequencies of less than 5 they must be combined with other cells until the restriction is satisfied. The $(r-1)(c-1)$ degrees of freedom should then be reduced by the number of cells that were combined with other cells.

Procedure for Testing a Hypothesis About a Contingency Table

(1) Formulate the null and alternative hypotheses. The same null hypothesis is always tested against the same alternative hypothesis:

H_0: The two criteria of classification are independent.
H_1: The two criteria of classification are *not* independent.

(2) Decide upon the α-level.

Look up χ_α^2; the number of degrees of freedom is $(r-1)(c-1)$. (See note below in this step about the degrees of freedom.) The critical region is $\chi^2 \geqslant \chi_\alpha^2$.

The test is always a one-sided test with the critical region on the right, because large values of χ^2 indicate poor agreement between observed and expected values.

NOTE: If some cells must be combined with others because they have expected frequencies that are less than 5 the $(r-1)(c-1)$ degrees of freedom must be decreased by the number of cells lost in the combining process. It is not known until step (3) whether any cells must be combined. If any are combined, and the number of degrees of freedom is different from $(r-1)(c-1)$, the correct value of χ_α^2 must be found after step (3) is performed.

(3) Compute the expected frequencies in each of the rc cells. The expected frequency of the (i, j)th cell (the cell in the ith row and jth column), where i takes on values from 1 to r and j takes on values from 1 to c, is

$$E_{ij} = \frac{R_i C_j}{n}$$

where R_i = total of the ith row;
C_j = total of the jth column; and
n = grand total.

If any theoretical frequency is less than 5, combine that cell with some other cell.

(4) Compute

$$\chi^2 = \sum \left\{ \frac{(O_{ij} - E_{ij})^2}{E_{ij}} \right\}$$

where O_{ij} = observed number in the (i, j)th cell; and
E_{ij} = expected (theoretical) number in (i, j)th cell.

The sum is to be taken over all the cells in the contingency table.

(5) Reject H_o if the value of χ^2 found in step (4) is in the critical region.

EXAMPLE: The members of a sample of size 89 were classified according to two criteria of classification. The results are shown in Table 11.12. Test the hypothesis that the two criteria of classification are independent.

Table 11.12

		Criterion 2		
		Level 1	Level 2	Totals
Criterion 1	Level 1	13	24	37
	Level 2	28	24	52
	TOTALS	41	48	89

(1) H_o: The criteria are independent.

H_1: The criteria are not independent.

(2) Let $\alpha = 0.05$, $r = 2$, and $c = 2$.

The number of degrees of freedom is $(r - 1)(c - 1) = (1)(1) = 1$.

$$\chi^2_{0.05}(1) = 3.841$$

(3)
$$R_1 = 37, R_2 = 52$$
$$C_1 = 41, C_2 = 48$$
$$n = 89$$
$$E_{11} = \frac{(37)(41)}{89} = 17.0$$
$$E_{12} = \frac{(37)(48)}{89} = 20.0$$
$$E_{21} = \frac{(52)(41)}{89} = 24.0$$
$$E_{22} = \frac{(52)(48)}{89} = 28.0$$

(4)
$$\chi^2 = \sum \left\{ \frac{(O_{ij} - E_{ij})^2}{E_{ij}} \right\}$$
$$= \frac{(13 - 17)^2}{17} + \frac{(24 - 20)^2}{20} + \frac{(28 - 24)^2}{24} + \frac{(24 - 28)^2}{28}$$
$$= \frac{16}{17} + \frac{16}{20} + \frac{16}{24} + \frac{16}{28}$$
$$= 0.941 + 0.8 + 0.667 + 0.571$$
$$= 2.98 \text{ correct to two decimal places}$$

(5)
$$2.98 < 3.841$$

Accept H_o.

A contingency table with r rows and c columns is known as an 'r by c table', written $r \times c$ table. The 2×2 table probably occurs more frequently than any other. (For instance, there is a test about the equality of the medians of two populations that makes use of 2×2 contingency tables.)

THE RANK-CORRELATION COEFFICIENT

The correlation coefficient was discussed in Chapter IX. If we take a random sample of n members and measure two characteristics of each member we obtain a random sample of n pairs of observations, $(x_1, y_1), (x_2, y_2), \ldots (x_n, y_n)$. Under the assumption that the distribution of x and y are normal the sample correlation coefficient is a measure of the linear relation between the two variables. There are two kinds of situations in which this approach cannot be used. In the first place, perhaps the variables cannot be assumed to have normal distributions. In the second, it might be impossible to measure the characteristics that are of interest.

The rank-correlation coefficient by-passes these difficulties very neatly because no normality assumption is required and, as the name indicates, it is based on ranks rather than on measurements. It is often easy to *rank* individuals with respect to some characteristic when it is difficult or impossible to *measure* that characteristic. The usual illustration of this remark is the beauty contest in which the contestants are ranked according to their beauty. Any sort of contest involving judges furnishes the same situation—a cooking contest, a talent contest, and so on.

The **Spearman rank-correlation** coefficient, which we will denote r_s, is the one that we will use (although it is not the only measure of correlation between ranks). If the values of x and y in the definition of the correlation coefficient, Formula (9.1), are replaced by their ranks, then that definition can be shown to yield

$$r_s = 1 - \frac{6 \sum (X_i - Y_i)^2}{n(n^2 - 1)} \qquad (11.4)$$

where r_s denotes the Spearman rank-correlation coefficient:

$n =$ number of pairs of observation;
$X_i =$ rank of x_i; and
$Y_i =$ rank of y_i.

The Spearman rank-correlation coefficient takes on values from -1 to $+1$, just as the correlation coefficient, r, does. A value of r_s equal to 1 indicates perfect agreement between the ranks of x and y. A value of r_s equal to -1 indicates that the ranks of y are in exactly the opposite order as the ranks of x. A value of r_s near zero indicates that x and y are independent.

Suppose there are two judges at a talent show that has six contestants. The first judge ranks the contestants 3, 4, 1, 2, 6, 5. The second judge also ranks the contestants 3, 4, 1, 2, 6, 5. Then the judges agree perfectly in their ranking, and the value of r_s will be 1.

On the other hand, suppose that the judge had ranked the six contestants 1, 2, 3, 4, 5, 6, and the second judge had ranked them 6, 5, 4, 3, 2, 1. The rankings of the judges are exactly reversed here—that is, if a contestant is ranked in the *j*th position by the first judge he is ranked in the $(n - j + 1)$th position by the second judge—and the value of r_s would be -1. Let us calculate r_s from Formula (11.4) for this last ranking.

Table 11.13

Contestant	X_i Judge 1	Y_i Judge 2	$X_i - Y_i$	$(X_i - Y_i)^2$
1	1	6	-5	25
2	2	5	-3	9
3	3	4	-1	1
4	4	3	1	1
5	5	2	3	9
6	6	1	5	25
				$70 = \sum (X_i - Y_i)^2$

We find $\sum (X_i - Y_i)^2 = 70$ and $n = 6$. Substituting these values into Formula (11.4), we have

$$r_s = 1 - \frac{6(70)}{6(35)} = 1 - 2 = -1$$

If measurements rather than ranks are given originally the measurements must be changed into ranks in order to find the rank-correlation coefficient. For example, suppose that the observations in Table 11.14 are the scores that 8 salesmen made on a test that measures their aggressiveness (x), and their sales in thousands of dollars for their second year with a certain company (y).

Table 11.14

Salesman	x	y
1	30	35
2	17	31
3	35	40
4	28	46
5	42	50
6	25	32
7	19	33
8	34	42

The x-values are ranked from largest to smallest (it does not matter whether the observations are ranked from largest to smallest, or from smallest to largest, as long as both x and y values are ranked the same way): 42, 35, 34, 30, 28, 25, 19, 17. The observation 42 is assigned rank 1, observation 35 is assigned rank 2, and so on. The y-values are ranked from largest to smallest also (they must be ranked the same way as the x-values): 50, 46, 42, 40, 35, 33, 32, 31. The observation 50 is assigned the rank 1, the observation 46 is assigned the rank 2, and so on.

Thus, Salesman 1 ranks fourth on the test score and fifth in sales: $X_1 = 4$ and $Y_1 = 5$. These two ranks appear in Row 1 of Table 11.15. The ranks of the test scores (X_i) of the eight salesmen and the ranks of their sales (Y_i) are shown in Table 11.15.

Table 11.15

Salesman	X_i	Y_i	$(X_i - Y_i)^2$
1	4	5	1
2	8	8	0
3	2	4	4
4	5	2	9
5	1	1	0
6	6	7	1
7	7	6	1
8	3	3	0

$$16 = \sum (X_i - Y_i)^2$$

The squared differences between ranks are shown in the last column of the table. Substituting the values $n = 8$ and $\sum (X_i - Y_i)^2 = 16$ into Formula (11.4), we obtain

$$r_s = 1 - \frac{6(16)}{8(63)} = 1 - 0 \cdot 1905 = 0 \cdot 8095 = 0 \cdot 810$$

The table of critical values of the Spearman rank-correlation coefficient in the Appendix is used to test the null hypothesis that the variables x and y (or, equivalently, their corresponding ranks, X and Y) are independent. The table gives the critical values of r_s for a one-sided alternative. When the alternative is two-sided the probabilities heading the column must be doubled to give the correct value of α.

For example, suppose that we test

H_0: x and y are independent

against $\qquad$ H_1: x and y are positively correlated

for a sample of size 14 and $\alpha = 0 \cdot 05$. We would reject H_0 if the sample yields a value of $r_s \geqslant 0 \cdot 456$.

On the other hand, if we want to test

H_0: x and y are independent

against the two-sided alternative

H_1: x and y are not independent

for $n = 14$ and $\alpha = 0 \cdot 10$, we reject H_0 if $r_s \geqslant 0 \cdot 456$ or if $r_s \leqslant -0 \cdot 456$.

EXAMPLE: For a sample of size 20 and $\alpha = 0 \cdot 01$, determine the critical region for testing

H_0: x and y are independent

against $\qquad$ H_1: x and y are negatively correlated.

The critical region consists of all values of $r_s \leqslant -0 \cdot 534$.

Using the table in the Appendix, the only possible α-values for a one-sided test are $\alpha = 0 \cdot 01$ and $\alpha = 0 \cdot 05$; for a two-sided test the only possible values for α are $\alpha = 0 \cdot 02$ and $\alpha = 0 \cdot 10$.

Non-parametric Statistics

183

Procedure for Testing H_0: Ranks Are Independent

(1) Formulate the null and alternative hypotheses:

H_0: The two sets of ranks are independent
H_1: The two sets of ranks are not independent

(If a one-sided test is desired it must be specified whether the correlation is positive or negative.)

(2) Decide upon a value of α. For a two-sided alternative, we use $\alpha = 0.02$ or $\alpha = 0.10$ (for the table in this book). For a one-sided alternative, we use $\alpha = 0.01$ or $\alpha = 0.05$. Look up the critical value (or values) of r_s for the appropriate values of α and n.

(3) Obtain a sample of n individuals, and measure two characteristics for each individual: $(x_1, y_1), (x_2, y_2), \ldots (x_n, y_n)$.

Rank the measurements of each characteristic from largest to smallest. We thus obtain pairs of ranks instead of pairs of observations:

$$(X_1, Y_1), (X_2, Y_2), \ldots (X_n, Y_n)$$

(4) Calculate $r_s = 1 - \dfrac{6 \sum (X_i - Y_i)^2}{n(n^2 - 1)}$

(5) Reject H_0 if r_s is in the critical region.

EXAMPLE: Nine applicants for two identical positions had individual interviews with each of two interviewers (the interview with Interviewer A was separate from the one with B). Each interviewer ranked the nine applicants according to their suitability for the job (personality, experience, intelligence, etc.). The applicants and the ranks that they received from each interviewer are given in Table 11.16. Test whether the rankings of the interviewers are independent.

Table 11.16

Applicant	X_i Rank by Interviewer A	Y_i Rank by Interviewer B	$(X_i - Y_i)^2$
1	4	2	4
2	3	4	1
3	7	6	1
4	2	1	1
5	9	9	0
6	8	8	0
7	1	3	4
8	5	7	4
9	6	5	1

(1) The null hypothesis that we are testing is that the ranks assigned by Interviewer A are independent of the ranks assigned by Interviewer B. We hope that the two interviewers would agree fairly well with each other (if two trained interviewers should rank two sets of applicants significantly different, then the interview method would appear to be an unsatisfactory way to evaluate prospective employees—or perhaps the applicants responded very differently to the two interviewers). Therefore we choose the alternative hypothesis that there is agreement—there is positive correlation—between the sets of ranks.

Stated more formally and concisely, we will test

H_0: The two sets of ranks are independent

against $\qquad H_1$: The two sets of ranks are positively correlated

(2) Let $\alpha = 0.05$.

Our sample size is 9. The critical value, from the table, is 0.600. The critical region consists of all values of r_s greater than or equal to 0.600.

(3) The two sets of ranks are given.

(4) The squared differences $(X_i - Y_i)^2$, appear in the last column of Table 11.16. Summing the entries in this column, we find

$$\sum (X_i - Y_i)^2 = 16$$

Substituting $\sum (X_i - Y_i)^2 = 16$ and $n = 9$ into Formula (11.4), we find that

$$r_s = 1 - \frac{6(16)}{9(80)}$$
$$= 1 - 0.133$$
$$= 0.867$$

(5) $\qquad\qquad\qquad\qquad\qquad 0.867 > 0.600$

Our value of r_s is in the critical region. Reject H_o.

THE SIGN TEST (ONE POPULATION)

The sign test is used to test the hypothesis that the median of a population equals a specified number. It is one of the simplest statistical tests to apply, and only two very general assumptions must be satisfied in order to apply it. These assumptions are: (1) the population being sampled is a continuous population; and (2) the population has a median.

Suppose that we are testing H_o: $M = M_o$ versus H_1: $M < M_o$. (Parameters are usually denoted by Greek letters, but we denote the median by M.) If the median of the population from which we are sampling is really equal to M_o we expect about half of the observations to be greater than M_o, and about half of the observations to be less than M_o. That is, if the null hypothesis is true the probability that any given observation will exceed M_o is $\frac{1}{2}$. If we obtain, in our sample, very many observations that are less than M_o we will be inclined to accept the alternative hypothesis rather than the null hypothesis.

In applying the sign test the hypothesized median, M_o, is subtracted from each observation, and only the sign of the difference recorded. If the median is actually M_o we would expect to obtain about as many plus signs as minus signs. If our sample yields a small number of either kind of sign and a large number of the other we will tend to reject the null hypothesis. We need a criterion by which we can determine when the observed results are so different from what we would expect that we will reject the null hypothesis.

This criterion is furnished us by the binomial distribution (actually, by the normal approximation to the binomial). If the null hypothesis is true—if the population median equals M_o—then the probability that any particular observation will exceed M_o (or be less than M_o) is $\frac{1}{2}$. Therefore the probability that any given observation will yield a plus sign (or minus sign) is $\frac{1}{2}$. The number of plus signs (or minus signs) that are yielded by the n observations has a binomial distribution with parameters $\frac{1}{2}$ and n. The theoretical histogram of such a distribution is symmetric. Even for small values of n, therefore, the binomial distribution with parameters $\frac{1}{2}$ and n is satisfactorily approximated by the normal curve with mean $n/2$ and variance $n/4$ (see Chapter VI), even when n is small.

In order to find the approximate probability that the number of plus signs,

x, is less than or equal to k (any positive integer), we must find the area under the normal curve with mean $n/2$ and variance $n/4$ to the left of $x = k + \frac{1}{2}$. Symbolically,

$$P_B(\mathbf{x} \leqslant k) \cong P_N(\mathbf{x} < k + 1/2)$$
$$= P_N\left(\mathbf{z} < \frac{(k + 1/2) - n/2)}{\sqrt{n/4}}\right)$$

For a two-sided test we will reject H_o if we get either too few or too many plus signs. For the level of significance α we reject H_o whenever the number of plus signs that we get leads to a z-value that falls into the critical region. For a two-sided test, if k plus signs are observed, where $k < n/2$, calculate

$$z = \frac{(k + 1/2) - n/2}{\frac{\sqrt{n}}{2}}$$

and reject H_o if $z \leqslant -z_{\alpha/2}$; if $k \geqslant \frac{n}{2}$, calculate

$$z = \frac{(k - 1/2) - n/2}{\frac{\sqrt{n}}{2}}$$

and reject H_o if $z \geqslant z_{\alpha/2}$.

When the alternative hypothesis is H_1: $M < M_o$ is true, then it is quite likely that more than half of the observations will be less than M_o, and, as a consequence, we will obtain few plus signs. If we obtain too few plus signs we are going to reject H_o and accept H_1. Then if our alternative is H_1: $M < M_o$, and we observe k plus signs, we ask ourselves: what is the probability of obtaining k or fewer plus signs, assuming that the null hypothesis is true? If the probability is smaller than α, of course, we reject the null hypothesis. The probability of obtaining k or fewer plus signs is less than α if

$$z = \frac{(k + 1/2) - n/2}{\frac{\sqrt{n}}{2}}$$

is less than $-z_\alpha$. Therefore, we simply calculate the z and reject if z is in the critical region.

If we are testing the one-sided alternative H_1: $M > M_o$, and if this alternative is true, we expect more than half of the observations in the sample to exceed M_o. Consequently, after subtracting M_o from each observation, and recording the signs of the differences, there should be more plus signs than minus signs. If we obtain too many plus signs we are going to reject H_o and accept H_1. Thus, when the alternative is H_1: $M > M_o$ and we obtain k plus signs we ask ourselves: what is the probability of obtaining k *or more* plus signs? The null hypothesis is rejected if this probability is less than α, and this occurs if

$$z = \frac{(k - 1/2) - n/2}{\frac{\sqrt{n}}{2}}$$

exceeds z_α. We reject H_o: $M = M_o$ and accept H_1: $M > M_o$ when $z \geqslant z_\alpha$.

Procedure for Testing H_0: $M = M_0$ versus H_1: $M \neq M_0$

(1) Formulate the null and alternative hypotheses:

$$H_0: M = M_0 \quad \text{versus} \quad H_1: M \neq M_0$$

(2) Decide upon a level of significance, α.

Find $z_{\alpha/2}$ in standard normal tables.

The critical region consists of those values of z less than or equal to $-z_{\alpha/2}$ and of those greater than or equal to $z_{\alpha/2}$.

(3) Select a random sample from the population.

Subtract M_0 from each observation. Record a plus sign for each positive difference and a minus sign for each negative difference. Discard any zero differences. Denote the number of non-zero differences (the total number of plus and minus signs) by n.

Denote the number of plus signs by k.

(4) Calculate z as follows:

If $k < n/2$, calculate $z = \dfrac{(k + 1/2) - n/2}{\dfrac{\sqrt{n}}{2}}$;

If $k \geqslant n/2$, calculate $z = \dfrac{(k - 1/2) - n/2}{\dfrac{\sqrt{n}}{2}}$

(5) If z is in the critical region, reject H_0 and accept H_1.

Procedure for Testing H_0: $M = M_0$ versus H_1: $M < M_0$

(1) Formulate the null and alternative hypotheses:

$$H_0: M = M_0 \quad \text{and} \quad H_1: M < M_0$$

(2) Decide upon a level of significance, α.

Find z_α in standard normal tables.

The critical region consists of all z-values that are less than or equal to $-z_\alpha$.

(3) Same as in testing H_0: $M = M_0$ versus H_1: $M \neq M_0$.

(4) Calculate

$$z = \frac{(k + 1/2) - n/2}{\dfrac{\sqrt{n}}{2}}$$

(5) Reject H_0 and accept H_1 if z is in the critical region.

Procedure for Testing H_0: $M = M_0$ versus H_1: $M > M_0$

(1) Formulate the null and alternative hypotheses. H_0: $M = M_0$ and H_1: $M > M_0$.

(2) Decide upon a level of significance, α.

Find z_α in standard normal tables.

The critical region consists of all those z-values that are greater than or equal to z_α.

(3) Same as in testing H_0: $M = M_0$ versus H_1: $M \neq M_0$.

(4) Calculate $z = \dfrac{(k - 1/2) - n/2}{\dfrac{\sqrt{n}}{2}}$

(5) Reject H_o and accept H_1 if z is in the critical region.

THE WILCOXON SIGNED-RANK TEST

The sign test discussed in the previous section does not require many assumptions and is easy to apply, but there is also an important disadvantage. If the population can be assumed to be symmetric, then the sign test does not use all of the information contained in the sample observations. (Incidentally, the mean and the median of a symmetric population have the same value (see Fig. 3.1).) For instance, common sense indicates that after the hypothetical median has been subtracted from each observation a difference of 10 should have more weight than a difference of 1. The sign test, however, treats these two differences alike, replacing each by $+$.

The Wilcoxon signed-rank test makes use of the relative sizes of the differences. The necessary assumptions are: (1) the population is continuous; (2) the population has a median; and (3) the population is symmetric. The hypothesized median, M_o, is subtracted from each observation, and the resulting differences are arranged in order of increasing absolute value (as in the sign test, any zero differences are discarded). The observations are then replaced by the ranks of the positions that they occupy, with observations that have the same absolute value being replaced by the mean (average) of the ranks that they occupy. The ranks of the positive differences (or negative differences, depending upon the alternative hypothesis) are added, the sum referred to the signed-rank table, and the null hypothesis accepted or rejected.

We will denote the smaller of the two rank-sums by T, and call it the **rank-sum statistic**. The sum of the ranks corresponding to positive differences will be called simply the positive rank-sum, and the sum of ranks corresponding to negative differences the negative rank-sum.

The Wilcoxon signed-rank table gives the critical value of T, the signed-rank statistic, for several values of α: $\alpha = 0.005$, 0.01, 0.025, and 0.05. If, in the testing of a hypothesis about a median, a value of T as small or smaller than the critical value is obtained the null hypothesis is rejected.

Suppose that we are testing the null hypothesis $H_o: M = M_o$ against the one-sided alternative $H_1: M < M_o$. The alternative hypothesis will be accepted if there are many more observations less than M_o than there are greater than M_o. If we do observe a great many observations less than M_o this will result in more negative differences than positive differences. As a consequence of the number of positive differences being small, we conclude that the positive rank-sum will also (presumably) be small. To repeat the previous reasoning in reverse order: if the positive rank-sum is small the number of observations greater than the hypothetical median is small (since the population is symmetric, we expect the observations to be scattered as much on one side of the median as the other), which indicates that the median is likely smaller than M_o. *Small values of the positive rank-sum lead to acceptance of $H_1: M < M_o$.*

The reasoning is similar when we are testing $H_o: M = M_o$ versus $H_1: M > M_o$. Small values of the negative rank-sum indicate a small number of negative differences (because of the symmetry assumption), which implies a small number of observations less than the hypothetical median, M_0, which implies that the median is larger than M_0. *Small values of the negative rank-sum lead to acceptance of H_1: $M > M_0$.*

When we are testing H_0: $M = M_0$ against the two-sided alternative, H_1: $M \neq M_0$, we are going to reject H_0 if the observed data seem to indicate that the median is actually greater than M_0, or if they indicate that it is actually less than M_0. If the median actually is very different from M_0 there will be considerably fewer observations on one side of the hypothetical median, M_0, than on the other, and therefore one of the rank-sums will be small. *Small values of either rank-sum lead to acceptance of the alternative hypothesis H_1: $M \neq M_0$.*

When a one-sided alternative is being tested, simply read the critical value under the appropriate value of α. For example, for a sample of size 13 (after discarding the zero differences) and $\alpha = 0.05$, we will reject H_0 in a one-sided test if T is less than or equal to 21; for $n = 10$ and $\alpha = 0.01$, we would reject H_0 if T is less than or equal to 5.

When a two-sided alternative is being tested, however, and the level of significance is α we read the critical value in the $\alpha/2$ column. That is, if $\alpha = 0.05$ we read the critical value in the column headed 0.025; if $\alpha = 0.01$ we read the critical value in the column headed 0.005. The reason for this is that when we test a two-sided alternative we choose whichever rank-sum is smaller to make the test. Our choosing the smaller rank-sum to compare with the critical value doubles the probability (shown at the top of the columns) of rejecting the null hypothesis. To compensate for this we must use the column for which the α-value is only half as great as the one we desire. Thus, for a two-sided alternative our reading in the 0.005 column really gives us an α-value of 0.01, and our reading in the 0.025 column really gives us an α-value of 0.05.

Procedure for Testing H_0: $M = M_0$ versus H_1: $M \neq M_0$

(1) Formulate the null and alternative hypotheses:

$$H_0: M = M_0 \quad \text{and} \quad H_1: M \neq M_0$$

(2) Select a random sample.

(3) Subtract M_0 from each observation, and record the differences.

Discard any zero differences, and rank the non-zero differences in order of increasing absolute value. Let the number of non-zero differences be denoted by n. Assign the ranks 1, 2, ... n to the ranked differences. If there are any differences that have the same absolute value, assign to each of these the average of the ranks that they occupy.

(4) Find the negative rank-sum and the positive rank-sum. Denote the smaller of these by T.

(5) In the table for the Wilcoxon signed-rank test, find, in the column headed by the value $\alpha/2$, the critical value of T for the appropriate value of n. Reject H_0 and accept H_1 if T is less than or equal to the critical value.

Procedure for Testing H_0: $M = M_0$ versus H_1: $M < M_0$

(1) Formulate the null and alternative hypotheses: H_0: $M = M_0$ and H_1: $M < M_0$.

Decide upon a value for α.

(2) Select a random sample.

(3) Same as in testing H_o: $M = M_o$ versus H_1: $M \neq M_o$.

(4) Find the positive rank-sum, and denote it by T.

(5) In the table for the Wilcoxon signed-rank test, find the critical value of T for the appropriate value of n in the column headed by the value α.

Reject H_o and accept H_1 if T is less than or equal to the critical value.

Procedure for Testing H_o: $M = M_o$ versus H_1: $M > M_o$

(1) Formulate the null and alternative hypotheses: H_o: $M = M_o$ and H_1: $M > M_o$.

    Decide upon a value for α.

(2) Select a random sample.

(3) Same as in testing H_o: $M = M_o$ versus H_1: $M \neq M_o$.

(4) Find the negative rank-sum, and denote it by T.

(5) In the table for the Wilcoxon signed-rank test, find the critical value of T for the appropriate value of n in the column headed by the value α.

Reject H_o and accept H_1 if T is less than or equal to the critical value.

EXAMPLE: Given that the eighteen observations below are a random sample from a continuous, symmetric population, test the null hypothesis that the median is 16 against the alternative that the median is not 16:

$$9, 11, 18, 16, 17, 21, 12, 10, 11, 11, 19, 16, 12, 13, 20, 14, 15, 13$$

(1) We are testing H_o: $M = 16$ versus H_1: $M \neq 16$. Let $\alpha = 0.05$.

(2) The random sample is given.

(3) Subtracting 16 from each observation, we get -7, -5, 2, 0, 1, 5, -4, -6. -5, -5, 3, 0, -4, -3, 4, -2, -1, -3.

Discarding the zeros and ranking the others in order of increasing absolute magnitude, we have 1, -1, 2, -2. 3, -3, -3, -4, -4, 4, -5, 5, -5, -5, -6, -7.

The 1s occupy ranks 1 and 2; the mean (average) of these ranks is 1.5; each 1 is given a rank of 1.5.

The 2s occupy ranks 3 and 4; the mean of these ranks is 3.5; each 2 is given a rank of 3.5.

In a similar manner, each 3 receives a rank of 6; each 4, a rank of 9; each 5, a rank of 12.5; the 6 is assigned a rank of 15; and the -7, a rank of 16.

The sequence of ranks is now 1.5, $\underline{1.5}$, 3.5, $\underline{3.5}$, 6, $\underline{6}$, $\underline{6}$, $\underline{9}$, $\underline{9}$, 9, $\underline{12.5}$, 12.5, $\underline{12.5}$, $\underline{12.5}$, $\underline{15}$, $\underline{16}$, with the 'negative ranks' underlined.

(4) The positive rank sum is 32.5.

    The negative rank sum is 103.5.

    $T = 32.5$

(5) $n = 16$

Since $\alpha = 0.05$, we read in the 0.025 column, and find the critical value to be 29. Accept H_o, because 32.5 is not less than or equal to 29.

THE RANK-SUM TEST (TWO POPULATIONS)

The rank-sum test is used to test the null hypothesis that the medians of two populations are equal against a specified alternative. The two populations must be assumed to be continuous and to have the same form, whatever it may be. If the two medians are equal as well, then the two samples are from two identical populations. It is not necessary that the samples contain the same number of observations. Designate the smaller sample as 'Sample 1' and the larger as 'Sample 2'. (If the samples are the same size it does not matter which sample is designated as Sample 1.) Let n_1 denote the number of observations in Sample 1, and n_2 the number of observations in Sample 2. The samples are combined, and the observations in the combined samples arranged in increasing order. The n_1 observations from Sample 1 are distinguished in some way. Underlining these observations is a good way to distinguish them, and is the method that we will use. Now the observations are replaced by their ranks, with equal observations being replaced by the mean of the ranks occupied by those observations. The ranks corresponding to the observations from Sample 1 are distinguished. The sum of the ranks occupied by the observations from Sample 1 is denoted by R.

If the assumptions are satisfied (if the observations are drawn from two identical continuous populations) the distribution of R can be found for various values of n_1 and n_2. The critical values of R are given for values of n_1 and n_2 as large as 10 and 10 in the table in the Appendix. The values of α are approximately 0·05 and approximately 0·025. For example, for sample of size 5 and 8, the probability that R will be less than or equal to 21 is 0·023; the probability that R will be greater than or equal to 49 is 0·023. The probability that R will be either less than or equal to 21 or greater than or equal to 49 is $2(0·023) = 0·046$. The probability that R will be less than or equal to 23 is 0·047; the probability that R will be greater than or equal to 47 is 0·047. The probability that R is either less than or equal to 23 or greater than or equal to 47 is $2(0·047) = 0·094$. Thus, if we want to test H_o: $M_1 = M_2$ against a two-sided alternative with $\alpha = 0·05$ (samples of sizes 5 and 8), the critical region consists of values of R less than or equal to 21 and of values of R greater than or equal to 49, and $\alpha = 2(0·023) = 0·046$.

For a one-sided alternative we need to select the proper critical value for R—the upper value or the lower value in the table. If the alternative H_1: $M_1 < M_2$ is true, then we expect R to be less than it would be if the null hypothesis, H_o: $M_1 = M_2$, were true. Therefore, small values of R lead us to reject H_o and accept H_1. For sample sizes of 5 and 8, if we want an approximate 5 per cent test of H_o: $M_1 = M_2$ versus H_1: $M_1 < M_2$, then the critical region consists of all values of R less than or equal to 23; $\alpha = 0·047$.

If the median of Population 1 is larger than that of Population 2 we expect the observations from Sample 1 to be larger than they would be if the two medians were equal and, hence, R, the sum of the ranks of those observations, will be larger than it would be if the medians were equal. Therefore, large values of R lead us to reject the null hypothesis and to accept the alternative, H_1: $M_1 > M_2$. For sample sizes of 5 and 8 and an α of about 0·05, the critical region for the test H_o: $M_1 = M_2$ versus H_1: $M_1 > M_2$ consists of all values of R greater than or equal to 47; $\alpha = 0·047$.

(The same test with the same underlying theory is sometimes presented under the name 'the modified rank-sum test'. The statistic used to make the test is not the rank-sum statistic, but the modified rank-sum statistic, and therefore the tables look different.)

Procedure for Testing H_o: $M_1 = M_2$ versus a Specified Alternative

(1) Formulate the null and alternative hypotheses. Decide upon the approximate value of α (it must be either 0·05 or 0·10).

(2) Draw a random sample from each of the two populations. The smaller sample is Sample 1 and contains n_1 observations; the larger sample is Sample 2

and has n_2 observations. (If the two samples are the same size it does not matter which one is designated as Sample 1.)

(3) Combine the two samples, ranking the observations in order from smallest to largest. Underline the observations from Sample 1. Replace the observations by their ranks (the smallest observation has rank 1, etc.).

If two or more observations are the same size they should be replaced by the mean of the ranks that they occupy.

Underline the ranks of the observations from Sample 1.

(4) Find R, the sum of the underlined ranks.

(5) Refer R to the table of rank-sum critical values. Reject H_o and accept H_1 if R is in the critical region. (See the prior material in this section if an explanation of how to use the table is desired.)

EXAMPLE: The two samples below are from two populations that are identical (except, perhaps, for their medians). Test the null hypothesis, H_o: $M_1 = M_2$ versus the alternative hypothesis, H_1: $M_1 > M_2$.

Sample 1: 44, 41, 47, 30, 36, 34, 47, 42.
Sample 2: 26, 38, 33, 35, 44, 25, 15, 48, 30, 32.

(1) We are testing H_o: $M_1 = M_2$ versus H_1: $M_1 > M_2$. Let $\alpha = 0.05$ (approximately).

(2) The random samples are given: $n_1 = 8$, $n_2 = 10$.

(3) Combining the two samples, we have

44, 41, 47, 30, 36, 34, 47, 42, / 26, 38, 33, 35, 44, 25, 15, 48, 30, 32

with the slant line (/) indicating where Sample 1 ends and Sample 2 begins. Arranging the observations in order from smallest to largest, we have

15, 25, 26, 30, 30, 32, 33, 34, 35, 36, 38, 41, 42, 44, 44, 47, 47, 48

Replacing the observations by their ranks, with equal observations being replaced by the mean of the ranks that they occupy, we obtain

1, 2, 3, 4·5, 4·5, 6, 7, 8, 9, 10, 11, 12, 13, 14·5, 14·5, 16·5, 16·5, 18

(4) The sum of the underlined ranks is 95.

(5) For $n_1 = 8$, $n_2 = 10$, and $\alpha = 0.05$ (approximately), the critical region consists of all values of R greater than or equal to 95; $\alpha = 0.051$.

$R = 95$. Reject H_o and accept H_1.

If neither sample contains more than 10 observations, and if the desired α is 0·05 or 0·10 (two-sided tests), or 0·025 or 0·05 (one-sided tests), the table of critical values for the rank-sum test is adequate. If either sample size (or both of them) exceeds 10 we can use a normal approximation to test the hypothesis.

It can be shown that the random variable R is approximately normal with mean

$$\mu_R = \frac{n_1(n_1 + n_2 + 1)}{2} \tag{11.5}$$

and variance

$$\sigma_R^2 = \frac{n_1 n_2(n_1 + n_2 + 1)}{12} \tag{11.6}$$

The random variable **R** is standardized. The value obtained in a particular experiment is referred to standard normal tables and the null hypothesis accepted or rejected.

If the alternative is two-sided, H_1: $M_1 \neq M_2$, then either very small or very large values of R will lead us to reject H_o: $M_1 = M_2$ and accept H_1. When the one-sided alternative, H_1: $M_1 < M_2$, is being tested small values of R lead us to accept the alternative hypothesis rather than the null. When the one-sided alternative, H_1: $M_1 > M_2$, is being tested large values of R lead us to accept the alternative hypothesis.

Procedure for Testing H_0: $M_1 = M_2$ versus H_1: $M_1 \neq M_2$ (At Least One Sample Size > 10)

(1) Furmulate the null and alternative hypotheses: H_0: $M_1 = M_2$ and H_1: $M_1 \neq M_2$.

(2) Decide upon a value for α.

Look up $z_{\alpha/2}$ in standard normal tables.

The critical region consists of all values of z (found in step 4) that are less than or equal to $-z_{\alpha/2}$ and of all values of z greater than or equal to $z_{\alpha/2}$.

(3) Draw the two random samples.

Combine the two samples, ranking the observations in order from smallest to largest. Underline the observations from Sample 1, the smaller sample.

Replace the observations by their ranks (the smallest observation has rank 1, etc.).

If two or more observations are the same size they should be replaced by the mean of the ranks that they occupy. Underline the ranks of the observations from Sample 1.

Find R, the sum of the underlined ranks.

(4) Calculate μ_R from Formula (11.5).

Calculate σ_R^2 from Formula (11.6). Calculate σ_R.

Calculate $z = \dfrac{R - \mu_R}{\sigma_R}$

(5) Reject H_0 if z is in the critical region.

Accept H_0 otherwise.

Procedure for Testing H_0: $M_1 = M_2$ versus H_1: $M_1 < M_2$ (Large Samples)

Same as that for testing H_0: $M_1 = M_2$ versus H_1: $M_1 \neq M_2$ except for step (2), which must be changed to read:

(2) Decide upon a value for α.

Look up z_α in standard normal tables.

The critical region consists of all values of z that are less than or equal to $-z_\alpha$.

Procedure for Testing H_0: $M_1 = M_2$ versus H_1: $M_1 > M_2$ (Large Samples)

Same as that for testing H_0: $M_1 = M_2$ versus H_1: $M_1 \neq M_2$, except for step (2), which must be changed to read:

(2) Decide upon a value for α.

Look up z_α in standard normal tables.

The critical region consists of all values of z that are greater than or equal to z_α.

EXAMPLE: Given the two samples below, test that the population medians are equal versus the alternative that $M_1 < M_2$.

Sample 1: 26, 25, 38, 33, 42, 40, 44, 26, 25, 43, 35, 48, 37.

Sample 2: 44, 30, 34, 47, 35, 46, 35, 47, 48, 34, 32, 42, 43. 49, 46, 47.

(1) We are testing $H_0: M_1 = M_2$ versus $H_1: M_1 < M_2$.

(2) Let
$$\alpha = 0 \cdot 05$$
$$z_{0.05} = 1 \cdot 645$$

The critical region consists of all values of z less than or equal to $-1 \cdot 645$.

(3) Combining the two samples, ranking in order from smallest to largest, and underlying observations from Sample 1, we get

$$\underline{25}, \ \underline{25}, \ \underline{26}, \ \underline{26}, \ 30, \ 32, \ \underline{33}, \ 34, \ 34, \ 35, \ 35, \ \underline{35}, \ \underline{37}, \ \underline{38}, \ \underline{40}, \ 42,$$
$$\underline{42}, \ 43, \ \underline{43}, \ 44, \ \underline{44}, \ 46, \ 46, \ 47, \ 47, \ 47, \ \underline{48}, \ 48, \ 49$$

There is no need to replace equal observations by the mean of the ranks occupied by those observations if the equal observations are in the same sample. For example, the observation 26 occupies ranks 3 and 4 above. The mean of these ranks is $3 \cdot 5$. Whether the two ranks are left as 3 and 4, or whether they are changed to $3 \cdot 5$ and $3 \cdot 5$ makes no difference—they still contribute a total of 7 to R. On the other hand, the observation 35 occupies ranks 10, 11, and 12. The mean of these ranks is 11. If the equal observations are not replaced by the mean of the ranks, the underlined 35 will contribute 12 to R, if they are replaced by the mean rank the 35 will contribute 11 to R. In summary, when equal observations have come from both populations they *must* be replaced by the mean of the ranks that they occupy; when the equal observations are all from the same population they *may* be replaced by the mean of the ranks.

Applying these comments, the observations replaced by their ranks are: $\underline{1}, \underline{2}, \underline{3},$ 4, 5, 6, $\underline{7}$, 8, 9, 11, 11, $\underline{11}$, $\underline{13}$, $\underline{14}$, $\underline{15}$, 16·5, $\underline{16·5}$, 18·5, $\underline{18·5}$, 20·5, $\underline{20·5}$, 22, $\underline{23}$, $\underline{24}$, $\underline{25}$, 26, $\underline{27·5}$, $\underline{27·5}$, 29. $R = 153$.

(4)
$$n_1 = 13, \ n_2 = 16$$

Substituting these values of n_1 and n_2 into Formulae (11.5) and (11·6), we get

$$\mu_R = \frac{(13)(30)}{2} = 195$$

and
$$\sigma_R^2 = \frac{(13)(16)(30)}{12} = 520$$

then
$$\sigma_R = 22 \cdot 8$$
$$z = \frac{R - \mu_R}{\sigma_R} = \frac{153 - 195}{22 \cdot 8} = -1 \cdot 84$$

(5) $-1 \cdot 84 < -1 \cdot 645$. Reject H_0 and accept H_1.

EXERCISES

1. In order to test the hypothesis that an ordinary die, numbered from 1 to 6, is honest, roll the die 60 times and compare the observed frequencies with the expected frequencies by means of the chi-squared test.

2. A biological research worker, studying rats, believes that $\frac{1}{16}$ of the litters containing four rats should contain no males; $\frac{1}{4}$ of the four-rat litters should contain one male; $\frac{3}{8}$ should contain two males; $\frac{1}{4}$ should contain three males; and $\frac{1}{16}$ should contain four males. In 400 litters he observes 20 litters with no males, 110 containing one male, 155 containing two males, 80 containing three males, and 35 containing four males. Is the theory compatible with the observed values?

3. When observations are classified according to two criteria of classification the following contingency table results. Test whether the two criteria of classification are independent. Let $\alpha = 0 \cdot 01$.

Criterion 1

Criterion 2		(1)	(2)	(3)	Totals
	(1)	250	200	150	600
	(2)	150	100	150	400
Totals		400	300	300	1000

4. A random sample of 1000 adults was obtained, and each member of the sample classified according to his educational level and according to the number of books he had read within the past year. Test whether the educational level is independent of the number of books read in the past year.

Educational Level

		12 Years or less	More than 12 years but not more than 16	More than 16 years	Totals
	None	330	50	20	400
	1	50	100	50	200
Books	2	100	150	50	300
Read	3 or more	20	30	50	100
	Totals	500	330	170	1000

5. Test that the median of the population from which the data below have been obtained equals 55 versus the alternative that it is less. Use the sign test.

48, 51, 49, 53, 61, 59, 45, 52, 65, 47, 57, 58, 65, 56, 45, 49, 54, 63, 46, 57, 54, 53, 52, 45

6. Test that the median of the population from which the data below have been obtained equals 150 versus the alternative that it does not. Use the sign test.

171, 165, 147, 158, 131, 138, 145, 160, 135, 149, 142, 152, 144, 137, 156, 147

7. Use the data in Exercise 5 to test, by means of the Wilcoxon signed-rank test, that the median equals 55 versus the alternative that the median is less than 55.

8. Use the data in Exercise 6 to test, by means of the Wilcoxon signed-rank test, that $M = 150$ against the alternative that $M \neq 150$.

9. Sample 1 below is obtained from Population 1 and Sample 2 is obtained from Population 2. Test that $M_1 = M_2$ against the alternative that $M_1 > M_2$.

Sample 1: 38, 32, 43, 37, 45, 38
Sample 2: 27, 39, 30, 22, 32, 29, 25, 30

10. Sample 1 below is obtained from Population 1 and Sample 2 is obtained from Population 2. Test that $M_1 = M_2$ against the alternative that $M_1 < M_2$.

Sample 1: 89, 90, 92, 81, 76, 88, 85
Sample 2: 78, 93, 81, 87, 89, 71, 90, 96, 92, 82

CHAPTER XII

THE ANALYSIS OF VARIANCE

INTRODUCTION

The analysis of variance is one of the most powerful tools at the disposal of the statistical worker. Essentially, the analysis of variance is a technique that separates the variation that is present into independent components; then these components are analysed in order to test certain hypotheses.

The simplest situation in which the analysis-of-variance technique can be applied is the **one-way analysis of variance**, also called the **one-variable-of-classification analysis of variance**, treated in the next two sections.

The hypothesis that is tested by means of the analysis-of-variance technique is whether the means of several populations are equal. This is the third time that we have discussed tests about means. In Chapters VII and VIII we learned how to test hypotheses about the mean of a single population. Tests of hypotheses about the means of two populations were also discussed in Chapter VIII. And, now, in this chapter we will treat the testing of the hypothesis that the means of several populations are equal.

In a one-variable-of-classification situation each observation is classified into one sample or another on the basis of a single criterion—the population from which it came (or, in statistical terms, the 'treatment' that it received). The word 'treatment' is used somewhat as the word 'success' is used when we talk about the binomial distribution. The members of the samples have not necessarily received a treatment of some sort, although they may have. For example, in an agricultural experiment (a very simple one) a person might be testing four different kinds of fertilizer on a number of plots of beans. The yield of each plot would then be classified into one of four samples according to which treatment of fertilizer it received. But in another situation an instructor might be interested in testing whether there is any difference in the test scores of four different classes on a particular test. It is, perhaps, not quite accurate to say that we will classify the students according to which treatment they received. However, this is the terminology that is used. Keep in mind that when something is said to have received a 'treatment' this statement should not necessarily be interpreted literally.

This chapter also deals with the **two-way analysis of variance** (also called the analysis of variance for two variables of classification), which is a little more complicated than the one-way analysis. Each observation is classified according to two criteria of classification. We will discuss only the situation in which there is one observation for each combination of the two treatments.

To summarize, there is a certain amount of variation present in every group of samples. The analysis of variance is a technique that separates the total variation present into separate independent components that may be attributed to one source or another. These separate components of the variance are then analysed (hence the name, analysis of variance) in such a manner that the

195

hypothesis that the population means are equal can be tested. The assumptions that must be made are: (1) we have random samples from normal distributions; and (2) the normal populations all have equal variances. We will denote this common variance by σ^2.

ONE-WAY ANALYSIS OF VARIANCE

In a one-variable-of-classification situation there are k samples, one from each of k normal populations with common variance, σ^2. (The term 'common variance' means that each population has the same variance.) The hypothesis to be tested is that the means of the populations are equal. This hypothesis is tested by comparing two independent estimates of the common variance σ^2.

Consider the following situation. An experimenter wants to determine which of four diets should be fed to chickens being raised for market. Each of the four diets is fed to six chickens from age 2 weeks to age 8 weeks. The four samples of six observations each in Table 12.1 represent the weight in grammes gained by the chickens during the feeding period.

Table 12.1

	Sample 1	Sample 2	Sample 3	Sample 4
	37	49	33	41
	42	38	34	48
	45	40	40	40
	49	39	38	42
	50	50	47	38
	45	41	36	41
$\sum$	268	257	228	250
Mean	44·67	42·83	38·00	41·67
Variance	22·67	27·77	26·00	11·47

The experimenter wants to test that the means of the four populations from which the four samples are obtained are equal. He must assume: (1) that the populations are normal; and (2) that the populations have a common variance, σ^2. Sample 1 is from a population with mean μ_1 and variance σ^2. The sample variance, s_1^2, is an estimate of the population variance σ^2. Sample 2 is from a population with mean μ_2 and variance σ^2. The sample variance, s_2^2, is an estimate of the population variance σ^2. Similar statements can be made about Samples 3 and 4. Recall that in Chapter VIII, where we needed to assume that the variances of two populations were equal, we pooled the two sample variances to get an estimate of the common population. Here we follow exactly the same procedure. We have four separate estimates of the common variance, σ^2. A better estimate of σ^2 than any of the separate estimates is obtained by pooling the four sample variances.

The pooled estimate of σ^2 is denoted s_p^2, and is given by

$$s_p^2 = \frac{(n_1 - 1)s_1^2 + (n_2 - 1)s_2^2 + \ldots + (n_k - 1)s_k^2}{n - k}, \qquad (12.1)$$

where $\quad s_i^2 =$ the variance of the ith sample;

$\quad\quad\quad\quad n_i =$ number of observations in the ith sample;

$\quad\quad\quad\quad n =$ total number of observations; and

$\quad\quad\quad\quad k =$ number of samples.

When the number of observations in each sample is the same, then s_p^2 is simply the mean of the sample variances. This can be shown very easily, as follows: let r represent the number of observations that each sample contains. The total number of observations is the product of the number of samples and the number of observations in each sample, namely, kr. Symbolically,

$$n_i = r \quad \text{for} \quad i = 1, 2, \ldots k$$

and $\quad\quad\quad\quad\quad n = kr$

Substituting these values into Formula (12.1), we have

$$s_p^2 = \frac{(r-1)s_1^2 + (r-1)s_2^2 + \ldots + (r-1)s_k^2}{kr - k} \quad (12.2)$$

The quantity $r - 1$ can be factored out of each term in the numerator of Formula (12.2), and the quantity k can be factored out of each term of the denominator. Thus, Formula (12.2) may be written

$$s_p^2 = \frac{(r-1)[s_1^2 + s_2^2 + \ldots + s_k^2]}{k(r-1)} \quad (12.3)$$

which is algebraically equivalent to

$$s_p^2 = \frac{s_1^2 + s_2^2 + \ldots + s_k^2}{k} \quad (12.4)$$

This last formula is simply the mean of the k sample variances. When the sample sizes are equal Formula (12.4), which is a special case of (12.1), may be used to compute the pooled variance; when the sample sizes are unequal Formula (12.1) *must* be used.

The sample variances for the four samples shown in Table 12.1 computed in the usual way (see Chapter IV). We obtain $s_1^2 = 22 \cdot 67$, $s_2^2 = 27 \cdot 77$, $s_3^2 = 26 \cdot 00$, $s_4^2 = 11 \cdot 47$. Our pooled estimate of the variance is

$$s_p^2 = \frac{22 \cdot 67 + 27 \cdot 77 + 26 \cdot 00 + 11 \cdot 47}{4} = 21 \cdot 98$$

We have assumed only that the populations have a common variance σ^2 in order to obtain the above estimate of the common variance. We have made no assumption whatsoever about the population means. So, no matter whether the means are equal or not, a valid estimate of the common variance, σ^2, is obtained by pooling the variances of the several samples.

Now we will obtain a second estimate of the common variance, σ^2, assuming that the population means are equal (as well as that the variances are equal). If the means *are* equal, then the estimate of σ^2 that we obtain (by assuming that they are) will be a valid one, and it should have a value that is reasonably close to that of the first estimate of σ^2, s_p^2. If the means are *not* equal, then the estimate of σ^2 that we obtain (by assuming that they are) will *not* be a valid one. We would expect the two estimates of σ^2 to differ, and if they differ significantly we will conclude that the assumption that the means are equal is false.

Assume, then, that the population means are equal:

$$\mu_1 = \mu_2 = \ldots = \mu_k = \mu$$

We therefore have k normal populations with a common mean μ as well as a common variance σ^2. If we have one sample from each of k identical populations this is equivalent to having k samples from the same population.

When we have k samples of the same size from the same population the k sample means can themselves be considered as a sample of k observations— *observations from the population of sample means*.

We now need to recall a previous statement. It was stated in Chapter VII that if $\bar{x}$ is based on a sample of size n from a normal distribution with mean μ and variance σ^2, then $\bar{x}$ has a distribution that is normal with mean μ and variance σ^2/n. Therefore, if we draw random samples of size r from a distribution with mean μ and variance σ^2 (or from k populations with a common variance σ^2 and a common mean μ) the sample means will have a distribution with mean μ and variance σ^2/r. That is, we can consider the means of the k samples, $\bar{x}_1, \bar{x}_2, \ldots \bar{x}_k$ to a random sample of k observations from a population whose variance is σ^2/r.

Since σ^2/r is the variance of the population of $\bar{x}$, we will denote it $\sigma_{\bar{x}}^2$. The variance of the sample of sample means will be denoted by $s_{\bar{x}}^2$. From

$$\sigma_{\bar{x}}^2 = \sigma^2/r$$

it follows that

$$\sigma^2 = r\sigma_{\bar{x}}^2$$

Therefore, if $s_{\bar{x}}^2$ is an estimate of $\sigma_{\bar{x}}^2$, then an estimate of σ^2 would be $rs_{\bar{x}}^2$. This estimate will be denoted s_c^2 (a c subscript is used because later this quantity will be referred to as the 'between columns mean square'). Thus, we have a second estimate of the common σ^2, namely,

$$s_c^2 = rs_{\bar{x}}^2 \tag{12.5}$$

This estimate has been obtained after making the assumption that the means of the k populations are equal.

If the population means are *not* equal, then the k sample means cannot be considered as a sample of k observations from a population of sample means. If the population means are not equal, then we would certainly expect more variability among the sample means than when the population means are equal. Thus, if the assumption of equal means is not satisfied we expect the variance of the sample of sample means, $s_{\bar{x}}^2$, to be larger than it would be if the assumption were true.

In order to test the hypothesis $H_0: \mu_1 = \mu_2 = \ldots = \mu_k$ versus the alternative H_1: the means are not all equal, we compare the two estimates of the common variance σ^2. If we have k samples, each of size r, from k normal populations with common mean and common variance, then the ratio

$$\frac{s_c^2}{s_p^2} \tag{12.6}$$

has an F-distribution with $k - 1$ and $k(r - 1)$ degrees of freedom. If the value of the ratio is near 1, then it is reasonable to conclude that s_c^2, as well as s_p^2, is a valid estimate of σ^2. On the other hand, if the ratio is much larger than 1 this indicates that s_c^2 is estimating not σ^2, but a parameter that is larger than σ^2, which indicates that the means are not all equal. Therefore, if the ratio in Formula (12.6) is significantly large we reject

$$H_0: \mu_1 = \mu_2 = \ldots = \mu_k$$

and conclude that the means are significantly different. By 'significantly large' we mean that the value of the ratio in (12.6) equals or exceeds the upper α-point of the F-distribution with $k-1$ and $k(r-1)$ degrees of freedom. More concisely, we reject H_0 if

$$\frac{s_c^2}{s_p^2} \geqslant F_\alpha[k-1, k(r-1)]$$

Referring again to the data in Table 12.1, we find that $\bar{x}_1 = 44.67$, $\bar{x}_2 = 42.83$, $\bar{x}_3 = 38.00$, and $\bar{x}_4 = 41.67$. When we treat these four values as a sample of four observations from the population of sample means the computations for finding the sample variance $s_{\bar{x}}^2$ in the usual way are

$$s_{\bar{x}}^2 = (1/3)\left[7010.207 - \frac{(167.17)^2}{4}\right]$$
$$= (1/3)[7010.207 - 6986.452]$$
$$= \frac{23.755}{3} = 7.9183$$

The size of each sample is $r = 6$. Our estimate of σ^2, based on the assumption that the population means are equal, is found by substituting the values for r and $s_{\bar{x}}^2$ into Formula (12.5):

$$s_c^2 = 6(7.9813) = 47.51$$

The value of s_c^2 is compared with the value of s_p^2, which was found earlier to be 21.98. We expect the two values to be near one another if the assumption of equal means is true. If s_c^2 is much larger than s_p^2 this indicates that it is likely that the means are not equal. When s_c^2 is significantly larger than s_p^2 we conclude that the means are *not* equal.

For the data in Table 12.1 the ratio in Formula (12.6) is

$$\frac{s_c^2}{s_p^2} = \frac{47.51}{21.98} = 2.16$$

If the hypothesis $H_0: \mu_1 = \mu_2 = \ldots = \mu_k$ is true (if the assumption of equal means is true), then s_c^2/s_p^2 has an F-distribution. The degrees of freedom in this particular case are $k-1 = 4-1 = 3$, and $k(r-1) = 4(6-1) = 20$. From the F-table we find that the upper 0.05-point for (3, 20) degrees of freedom is 3.10. We accept the null hypothesis because the value 2.16 is not significantly large.

It is necessary to introduce here some general notation so that computing formulas for s_p^2 and s_c^2 can be written.

We need to place *two* subscripts on the x's. The first subscript, i, denotes the sample—1, 2, ... or k. The second subscript, j, denotes the observation within the sample. Each sample has r observations, so j might equal 1, 2, ... or r. Thus, the symbol x_{ij} denotes the observation in the jth row and ith column. In particular, x_{15} represents the fifth observation in the first sample; x_{34} means the fourth observation in the third sample; and so on. The display in Table 12.2 represents a general group of k samples, each containing r observations.

Table 12.2

Sample	1	2	...	k
	x_{11}	x_{21}	...	x_{k1}
	x_{12}	x_{22}	...	x_{k2}
	.	.	.	.
	.	.	.	.
	.	.	.	.
	x_{1r}	x_{2r}	...	x_{kr}
Totals	$T_1.$	$T_2.$	...	$T_k.$
Means	$\bar{x}_1$	$\bar{x}_2$	...	$\bar{x}_k$

The sum $\displaystyle\sum_{j=1}^{r} x_{1j}$ represents the total of the observations in the first sample. We will denote this total by $T_1.$ (read 'tee one dot'). The dot replaces the subscript over which we have summed. Thus,

$$\sum_{j=1}^{r} x_{1j} = T_1.$$

and the same notation is used for the sums of the observations in the other samples:

$$\sum_{j=1}^{r} x_{2j} = T_2.$$

$$\vdots$$

$$\sum_{j=1}^{r} x_{kj} = T_k.$$

In general, $T_i.$ denotes the sum of the observations in the ith sample (or column).

It can be shown that a convenient computing formula for s_p^2 is

$$s_p^2 = \frac{\displaystyle\sum_{j=1}^{r}\sum_{i=1}^{k} x_{ij}^2 - \frac{1}{r}(T_1.^2 + T_2.^2 + \ldots + T_k.^2)}{k(r-1)} \tag{12.7}$$

You are already familiar with the use of summation notation. Although the symbol $\displaystyle\sum_{j=1}^{r}\sum_{i=1}^{k} x_{ij}^2$ looks unfamiliar and difficult, it simply represents the sum of the squares of every observation in every sample. The double summation $\displaystyle\sum_{j=1}^{r}\sum_{i=1}^{k} x_{ij}^2$ can also be written $\displaystyle\sum_{i=1}^{k}\sum_{j=1}^{r} x_{ij}^2$ or $\displaystyle\sum_{j=1}^{r}\left(\sum_{i=1}^{k} x_{ij}^2\right)$ or $\displaystyle\sum_{i=1}^{k}\left(\sum_{j=1}^{r} x_{ij}^2\right)$.

Before we display a computing formula s_c^2 we need a symbol for the sum of all the observations in the k samples. Following the same notation that was used for the totals of the individual samples, this sum is denoted by $T..$ and is usually called the grand total. Symbolically,

$$T.. = T_1. + T_2. + \ldots + T_k. = \sum_{j=1}^{r} \sum_{i=1}^{k} x_{ij} \qquad (12.8)$$

Note that $\sum_{j=1}^{r} \sum_{i=1}^{k} x_{ij}$ is shorthand for the sum of all the observations.

It can be shown that a convenient computing formula for s_c^2 is

$$s_c^2 = \frac{\frac{1}{r}\left(\sum_{i=1}^{k} T_{i.}^2\right) - \frac{T_{..}^2}{rk}}{k-1} \qquad (12.9)$$

Note that the square of the grand total, $T_{..}^2$, is divided by the number of observations, kr, that were added to obtain the grand total. Also note that each of the squares of the separate sample totals is divided by the number of observations that were added together to obtain that total (r in every case— hence we can add all of the totals before dividing by r instead of dividing each total separately).

As an illustration of the use of these computing formulae we will verify the values of s_p^2 and s_c^2 obtained earlier in this section.

EXAMPLE: Use the computing formula in (12.7) to find the value of s_p^2 for the data in Table 12.1. (Recall that we found the value of s_p^2 by another method before a computing formula was available.)

First, we need to find the sum of all the observations:

$$\sum \sum x_{ij}^2 = (37)^2 + (42)^2 + \ldots + (38)^2 + (41)^2 = 42\ 499$$

Second, we must find the sum of the squares of the sample totals. The sample totals themselves have already been found: they are $T_1. = 268$, $T_2. = 257$, $T_3. = 228$, $T_4. = 250$. Therefore,

$$T_1^2. + T_2^2. + T_3^2. + T_4^2. = (268)^2 + (257)^2 + (228)^2 + (250)^2$$
$$= 71\ 824 + 66\ 049 + 51\ 984 + 62\ 500$$
$$= 252\ 357$$

Substituting these two values into Formula (12.7), along with the values $r = 6$ and $k = 4$, we obtain

$$s_p^2 = \frac{42\ 499 - \frac{1}{6}(252\ 357)}{4(6-1)} = \frac{439 \cdot 5}{20} = 21 \cdot 98$$

EXAMPLE: Use the computing formula in (12.9) to find the value of s_c^2 for the data in Table 12.1. (Recall again that we found the value of s_c^2 by another method before a computing formula was available.)

When we examine (12.9) we see that the necessary sums are $\sum_{i=1}^{k} T_i.^2$ and $T...$ In the previous example, we found

$$\sum_{i=1}^{k} T_i^2. = 252\ 357$$

We find $T..$ to be

$$T.. = \sum_{i=1}^{k} T_{i.} = (268) + (257) + (228) + (250)$$

$$= 1003$$

Substituting these two sums into Formula (12.9), along with the values $k = 4$ and $r = 6$, we obtain

$$s_c^2 = \frac{\dfrac{252\ 357}{6} - \dfrac{(1003)^2}{24}}{3}$$

$$= \frac{42\ 059 \cdot 5 - 41\ 917 \cdot 0}{3}$$

$$= \frac{142 \cdot 5}{3} = 47 \cdot 5$$

ONE-WAY ANALYSIS OF VARIANCE—ANOTHER APPROACH

The situation discussed in the last section will now be considered from another point of view. The situation was: we have a sample from each of k normal populations. Each sample contains the same number of observations, r. (The case in which the sample sizes are not all equal will be discussed later.)

The grand mean is the mean of all the observations in all the samples. We will denote the grand mean by the symbol $\bar{x}$ where

$$\bar{x} = \frac{T..}{rk} \tag{12.10}$$

where $T..$ is the grand total of all the observations.

The total variation present in the k samples is the sum of the deviations of the sample observations (in all samples) from the grand mean. This total variation is known as the **total sum of squares**. This will be denoted SST, and is given by

$$SST = \sum_{j=1}^{r} \sum_{i=1}^{k} (x_{ij} - \bar{x})^2 \tag{12.11}$$

We can derive an interesting result if we add and subtract the quantity $\bar{x}_i$ in Formula (12.11). Obviously,

$$x_{ij} - \bar{x} = x_{ij} - \bar{x}_i + \bar{x}_i - \bar{x}$$

and grouping the first two and last two terms of the expression on the right-hand side, we have

$$x_{ij} - \bar{x} = (x_{ij} - \bar{x}_i) + (\bar{x}_i - \bar{x})$$

If we replace $x_{ij} - \bar{x}$ in (12.11) by this equivalent quantity, we have

$$\sum \sum [x_{ij} - \bar{x}]^2 = \sum \sum [(x_{ij} - \bar{x}_i) + (\bar{x}_i - \bar{x})]^2 \tag{12.12}$$

Squaring the quantity in brackets, (12.12) becomes

$$\sum \sum (x_{ij} - \bar{x})^2 = \sum \sum [(x_{ij} - \bar{x}_i)^2 + 2(x_{ij} - \bar{x}_i)(\bar{x}_i - \bar{x}) + (\bar{x}_i - \bar{x})^2]$$

which, applying one of the properties of summations (see Chapter III), yields

$$\sum \sum (x_{ij} - \bar{x})^2 = \sum \sum (x_{ij} - \bar{x}_i)^2 + 2 \sum \sum (x_{ij} - \bar{x}_i)(\bar{x}_i - \bar{x}) + \sum \sum (\bar{x}_i - \bar{x})^2 \tag{12.13}$$

Now we will examine separately each of the three double summations on the right-hand side of Formula (12.13). First consider the double summation

$$\sum_{i=1}^{k}\sum_{j=1}^{r}(x_{ij}-\bar{x}_i)(\bar{x}_i-\bar{x})$$

The order of the two summation symbols indicates that we are to add over the subscript j first, and then add over the i subscript. We can use brackets to indicate this order, and write

$$\sum_{i=1}^{k}\sum_{j=1}^{r}(x_{ij}-\bar{x}_i)(\bar{x}_i-\bar{x})=\sum_{i=1}^{k}\left[\sum_{j=1}^{r}(x_{ij}-\bar{x}_i)(\bar{x}_i-\bar{x})\right]$$

When summing over the subscript j, any factor that does not contain a j is considered as a constant, and can be factored out of the summation. Therefore, the quantity $(\bar{x}_i-\bar{x})$ can be factored out of the summation within brackets, and we have

$$\sum_{i=1}^{k}\sum_{j=1}^{r}(x_{ij}-\bar{x}_i)(\bar{x}_i-\bar{x})=\sum_{i=1}^{k}\left[(\bar{x}_i-\bar{x})\cdot\sum_{j=1}^{r}(x_{ij}-\bar{x}_i)\right] \quad (12.14)$$

The summation $\sum_{j=1}^{r}(x_{ij}-\bar{x}_i)$ can be shown to be equal to zero. The steps are shown below:

$$\sum_{j=1}^{r}(x_{ij}-\bar{x}_i)=\sum_{j=1}^{r}x_{ij}-\sum_{j=1}^{r}\bar{x}_i \quad \text{(property of summations)}$$

$$=T_i.-\sum_{j=1}^{r}\bar{x}_i \quad \left(\sum_{j=1}^{r}x_{ij}=T_i.\right)$$

$$=T_i.-r\bar{x}_i \quad (\bar{x}_i \text{ is a constant when summed over } j)$$

$$=T_i.-r\frac{T_i.}{r} \quad \left(\bar{x}_i=\frac{T_i.}{r} \text{ by definition}\right)$$

$$=T_i.-T_i. \quad \text{algebra}$$

$$=0$$

Therefore Formula (12.14) becomes

$$\sum_{i=1}^{k}\sum_{j=1}^{r}(x_{ij}-\bar{x}_i)(\bar{x}_i-\bar{x})=\sum_{i=1}^{k}[(\bar{x}_i-\bar{x})\cdot 0]$$

$$=\sum_{i=1}^{k}[0]=0$$

and Formula (12.13) becomes

$$\sum_{i=1}^{k}\sum_{j=1}^{r}(x_{ij}-\bar{x})^2=\sum_{i=1}^{k}\sum_{j=1}^{r}(x_{ij}-\bar{x}_i)^2+\sum_{i=1}^{k}\sum_{j=1}^{r}(\bar{x}_i-\bar{x})^2 \quad (12.15)$$

All of the variation present in the samples can be separated into two pieces. The quantity

$$\sum\sum(x_{ij}-\bar{x}_i)^2$$

is called the **error sum of squares** or the **within (samples) sum of squares**, and is that part of the total variation that is due to the deviations of the observations from their particular sample means.

The quantity

$$\sum \sum (\bar{x}_i - \bar{x})^2$$

is called the **among sum of** squares, or, less grammatically, the **between sum of squares**. It is that part of the total variation that is due to the variation among the sample means; or, equivalently, due to the variation among samples; or, equivalently, since all of the observations in each sample have been written in the same column (see Tables 12.1 and 12.2), due to the variations among columns. Hence, the among sum of squares is sometimes called the **among columns sum of squares** if each column of the data contains the observations from one sample.

The computing formula for the error sum of squares, which will be denoted *SSE*, is

$$SSE = \sum_{i=1}^{k} \sum_{j=1}^{r} x_{ij}^2 - \sum_{i=1}^{k} \frac{T_{i\cdot}^2}{r} \qquad (12.16)$$

We use Formula (12.16) to calculate *SSE*, if we want to calculate *SSE* directly. (As we shall soon see, *SSE* is usually calculated indirectly.)

The computing formula for the among columns sum of squares, denoted *SSC*, is

$$SSC = \left(\sum_{i=1}^{k} \frac{T_{i\cdot}^2}{r} \right) - \frac{T_{\cdot\cdot}^2}{rk} \qquad (12.17)$$

The computing formula for the total sum of squares, which was defined in (12.11), is

$$SST = \left(\sum_{j=1}^{r} \sum_{i=1}^{k} x_{ij}^2 \right) - \frac{T_{\cdot\cdot}^2}{rk} \qquad (12.18)$$

We mentioned briefly earlier that the error sum of squares is usually calculated indirectly. Formula (12.15) gives the relationship

$$SST = SSE + SSC$$

Solving this equation for *SSE*, we obtain

$$SSE = SST - SSC \qquad (12.19)$$

In practice, *SST* and *SSC* are found, and then *SSE* is found by subtraction.

It is necessary to find the sums of squares because they are used to find quantities that are known as **mean squares** (you have already had experience with mean squares—the sample variance discussed in Chapter IV is a mean square). The mean squares are used to test the hypothesis that the population means are equal. We find mean squares for only two of the three above sums of squares; we do not find a mean square associated with the total sum of squares. In general, the mean square equals the sum of squares divided by the number of degrees of freedom associated with that sum of squares. This relationship can be stated as a formula:

$$\text{Mean square} = \frac{\text{Sum of squares}}{\text{Degrees of freedom}}$$

or more briefly,

$$M.S. = \frac{S.S.}{d.f.}$$

The number of degrees of freedom associated with the among columns sum of squares is one less than the number of columns. There are k columns; hence, there are $k - 1$ degrees of freedom. Thus, the among columns mean square, denoted MSC, is given by the formula

$$MSC = \frac{SSC}{k - 1} \tag{12.20}$$

When we replace SSC by its computing formula from (12.17) we obtain a computing formula for MSC,

$$MSC = \frac{\left(\sum_{i=1}^{k} \frac{T_{i \cdot}^2}{r}\right) - \frac{T_{\cdot \cdot}^2}{rk}}{k - 1} \tag{12.21}$$

Compare the expression given for MSC above and the expression given for s_c^2 in Formula (12.9). The two formulae are identical. Thus we see that MSC is the same as s_c^2, the estimate of the common variance, σ^2, that we obtained under the assumption that the population means were equal.

The degrees of freedom associated with the error sum of squares is, in general, equal to the sum of the k numbers obtained by subtracting 1 from each sample size. When each of the k samples contains r observations there are $k(r - 1)$ degrees of freedom. Then the error mean square, which we will denote MSE, is given by the formula

$$MSE = \frac{SSE}{k(r - 1)} \tag{12.22}$$

when the sample sizes are equal in the one-variable-of-classification analysis of variance. The degrees of freedom are different in other situations. When we replace SSE by its computing formula given in (12.16), then Formula (12.22) becomes

$$MSE = \frac{\sum \sum x_{ij}^2 - \sum_{i=1}^{k} \frac{T_{i \cdot}^2}{r}}{k(r - 1)} \tag{12.23}$$

When we compare the formula for MSE given above and the formula for s_p^2 given in (12.7) we see that the two expressions are equal. Therefore we can say that the MSE is the same quantity as s_p^2, which is an estimate of the common variance, σ^2, obtained *without* making the assumption that the population means are equal.

Therefore, in order to test the hypothesis

$$H_0: \mu_1 = \mu_2 = \ldots \mu_k$$

we form the ratio $$F = \frac{MSC}{MSE}$$

and if F exceeds the upper α-point of the F-distribution we reject the null hypothesis.

Usually a table known as an **analysis of variance table** is used to display the various sources of variation, along with the sums of squares, degrees of freedom, and mean squares associated with those sources. The analysis-of-variance table for a one-variable-of-classification analysis of variance, with k samples of r observations each, is shown in Table 12.3.

Table 12.3. Analysis of Variance Table

Source	d.f.	Sum of Squares	Mean square	F
Among columns	$k-1$	$SSC = \sum \frac{T_{i.}^2}{r} - \frac{T_{..}^2}{rk}$	$MSC = \frac{SSC}{k-1}$	$\frac{MSC}{MSE}$
Error (within)	$k(r-1)$	$SSE = SST - SSC$	$MSE = \frac{SSE}{k(r-1)}$	
Total	$rk-1$	$SST = \sum \sum x_{ij}^2 - \frac{T_{..}^2}{rk}$	——	

Procedure: One-way Analysis of Variance, Equal Sample Sizes

(1) Formulate the null and alternative hypotheses:

$$H_0: \mu_1 = \mu_2 = \ldots = \mu_k$$
and $\quad\quad H_1$: the means are not all equal

(2) Decide upon an α-level.
 Find $F_\alpha[k-1, k(r-1)]$ in the F-table.

(3) Obtain a random sample of size r from each of k normal populations. (Or, apply each of k 'treatments' to r 'members', assuming that the data so obtained are random samples from normal populations.)

(4) Calculate $T_{1.}, T_{2.}, \ldots T_k$.
 Calculate $T_{..}$.
 Calculate $\frac{T_{..}^2}{rk}$
 Calculate $\sum \sum x_{ij}^2$

(5) Calculate $SST = \sum \sum x_{ij}^2 - \frac{T_{..}^2}{rk}$

 Calculate $SSC = \left(\frac{T_{1.}^2}{r} + \frac{T_{2.}^2}{r} + \ldots + \frac{T_{k.}^2}{r} \right) - \frac{T_{..}^2}{rk}$

 Calculate $SSE = SST - SSC$

(6) Complete the analysis of variance table.

(7) Calculate $F = \frac{MSC}{MSE}$

(8) Reject H_0 if $F \geqslant F_\alpha[k-1, k(r-1)]$.

EXAMPLE: Given the data below, test the hypothesis that the means of the five populations are equal.

Sample 1	Sample 2	Sample 3	Sample 4	Sample 5
9	5	11	9	16
11	6	14	10	15
12	8	15	7	19
7	8	10	9	14

(1) We are testing $\quad H_0: \mu_1 = \mu_2 = \mu_3 = \mu_4 = \mu_5$
against the alternative $\quad H_1$: the means are not all equal

(2) Let $\alpha = 0.05$.
 We have $k = 5$, $r = 4$, and $F_{0.05}(4, 15) = 3.06$.

(3) The data are given.

(4)
$$T_{1.} = 9 + 11 + 12 + 7 = 39$$
$$T_{2.} = 5 + 6 + 8 + 8 = 27$$
$$T_{3.} = 11 + 14 + 15 + 10 = 50$$
$$T_{4.} = 9 + 10 + 7 + 9 = 35$$
$$T_{5.} = 16 + 15 + 19 + 14 = 64$$
$$T_{..} = 39 + 27 + 50 + 35 + 64 = 215$$

$$\frac{T_{..}^2}{rk} = \frac{(215)^2}{4.5} = \frac{46\,225}{20} = 2311 \cdot 25$$

$$\sum\sum x_{ij}^2 = (9)^2 + (11)^2 + (12)^2 + \ldots + (15)^2 + (19)^2 + (14)^2$$
$$= 81 + 121 + 144 + \ldots + 225 + 361 + 196 = 2575$$

(5)
$$SST = 2575 - 2311 \cdot 25 = 263 \cdot 75$$

$$SSC = \frac{(39)^2}{4} + \frac{(27)^2}{4} + \frac{(50)^2}{4} + \frac{(35)^2}{4} + \frac{(64)^2}{4} - \frac{(215)^2}{20}$$

$$= 380 \cdot 25 + 182 \cdot 25 + 625 \cdot 00 + 306 \cdot 25 + 1024 \cdot 00 - 2311 \cdot 25$$
$$= 2517 \cdot 75 - 2311 \cdot 25 = 206 \cdot 50$$

$$SSE = 263 \cdot 75 - 206 \cdot 50 = 57 \cdot 25$$

(6) The analysis of variance table:

Source	d.f.	Sum of squares	Mean square
Among columns	4	206·50	51·63
Error	15	57·25	3·82
Total	19	263·75	

(7)
$$F = \frac{MSC}{MSE} = \frac{51 \cdot 63}{3 \cdot 82} = 13 \cdot 52$$

(8) $13 \cdot 52 > 3 \cdot 06$. Reject H_0.

ONE-WAY ANALYSIS OF VARIANCE, DIFFERENT SAMPLE SIZES

In the previous two sections the one-way analysis of variance was discussed for the situation in which there are k samples with each sample containing the same number of observations, namely r. For various reasons there are many situations in which the sizes of the samples are not equal. One sample might contain ten observations, another five, and yet another thirteen. The computing formulae derived in the preceding section, for equal sample sizes, need only slight modifications in order to give the computing formulae that are applicable when the sample sizes are not equal.

In general, we will say that the first sample contains n_1 observations, the second sample contains n_2 observations, and so on, to the kth sample, which contains n_k observations. As before, we need two subscripts to denote each separate observation, the first indicating the sample and the second indicating the number of the observation within the sample. For example, x_{47} is the seventh observation in the fourth sample, and x_{2n_2} is the n_2th observation (the last one) in the second sample.

A general set of k samples, with Sample 1 containing n_1 observations,

Sample 2 containing n_2 observations, and so on, is shown in Table 12.4. The notation for the totals of the observations in the various samples is exactly the same as in the preceding section. That is, $T_1.$ is the total of the observations in the first sample; symbolically

$$T_1. = \sum_{j=1}^{n_1} x_{1j}.$$

Also, $T_2.$ is the total of the observations in the second sample; symbolically

$$T_2. = \sum_{j=1}^{n_2} x_{2j}.$$

In general, $T_i.$ is the total of the observations in the ith sample and is given by the formula

$$T_i. = \sum_{j=1}^{n_i} x_{ij}.$$

The appropriate total is shown at the bottom of each sample.

The grand total is the sum of all observations, and is denoted by $T..$, as before.

Table 12.4

	Sample 1	Sample 2	....	Sample k
	x_{11}	x_{21}		x_{k1}
	x_{12}	x_{22}		x_{k2}
	.	.		.
	.	.		.
	.	.		.
	.	.		.
	.	.		.
	x_{1n_1}	x_{2n_2}		x_{kn_k}
Totals	$T_1.$	$T_2.$		$T_k.$

As before, the hypothesis that is tested is that the k population means are equal. This hypothesis is tested by calculating the ratio of the among samples (columns) mean square to the within samples mean square, $\dfrac{MSC}{MSE}$, and rejecting the null hypothesis when this value exceeds the upper α-point of the F-distribution (with the appropriate degrees of freedom). In order to obtain MSC and MSE, we need to find the total sum of squares, the among columns sum of squares, and the within sum of squares.

In the last section the computing formula for the total sum of squares was

$$SST = \left(\sum_{i=1}^{k} \sum_{j=1}^{r} x_{ij}^2 \right) - \frac{T..^2}{rk} \qquad (12.24)$$

The quantity $T..$ was obtained by adding rk observations. In Formula (12.24) note that the square of the grand total, $T..^2$, is divided by rk, the number of observations that were added together to obtain the total. When the sample sizes are different the grand total, $T..$, is obtained by adding n observations, where

$$n = n_1 + n_2 + \ldots + n_k \qquad (12.25)$$

Therefore, the quantity $T_{..}^2/n$ is analogous to the quantity $T_{..}^2/rk$ that appears in Formula (12.24).

The only change in the quantity $\sum_{i=1}^{k}\sum_{j=1}^{r} x_{ij}^2$ that must be made when the sample sizes are different is in the limits of summation. The summation $\sum_{i=1}^{k}\sum_{j=1}^{m_i} x_{ij}^2$ is analogous to the summation $\sum_{i=1}^{k}\sum_{j=1}^{r} x_{ij}^2$, which appears in Formula (12.24). Both summations represent the sum of the squares of all of the observations.

Therefore, by analogy with the computing formula for the total sum of squares when each sample contains the same number of observations, given in (12.24), we can say that

$$SST = \left(\sum_{i=1}^{k}\sum_{j=1}^{n_i} x_{ij}^2\right) - \frac{T_{..}^2}{n} \tag{12.26}$$

is the computing formula for the total sum of squares when the sample sizes are not the same.

When the samples were all of the same size, r, the computing formula for the among columns sum of squares was given in Formula (12.17),

$$SSC = \left(\frac{T_{1.}^2}{r} + \frac{T_{2.}^2}{r} + \ldots + \frac{T_{k.}^2}{r}\right) - \frac{T_{..}^2}{rk} \tag{12.27}$$

As we have just seen, when the sample sizes are unequal the quantity $T_{..}^2/n$ is used instead of $T_{..}^2/rk$. Each of the column totals, $T_{i.}$, is obtained by adding r observations when the sample sizes are equal. In Formula (12.27) each of the squares of the column totals is divided by r, the number of observations that were added to obtain each total.

When the sample sizes are different the column total $T_{1.}$ is obtained by adding n_1 observations, the column total $T_{2.}$ is obtained by adding n_2 observations, and, in general, the column total $T_{i.}$ is obtained by adding n_i observations. Therefore, by analogy with the quantities in (12.27), and in light of the previous comments, $T_{1.}^2$ should be divided by n_1, $T_{2.}^2$ should be divided by n_2, and so on. When the sample sizes are different, instead of the quantity

$$\frac{T_{1.}^2}{r} + \frac{T_{2.}^2}{r} + \ldots + \frac{T_{k.}^2}{r}$$

we use the quantity $\quad \dfrac{T_{1.}^2}{n_1} + \dfrac{T_{2.}^2}{n_2} + \ldots + \dfrac{T_{k.}^2}{n_k}$

Therefore, the computing formula for SSC, when the sample sizes are different, is

$$SSC = \left(\frac{T_{1.}^2}{n_1} + \frac{T_{2.}^2}{n_2} + \ldots + \frac{T_{k.}^2}{n_k}\right) - \frac{T_{..}^2}{n} \tag{12.28}$$

Note the similarity between Formula (12.27) and (12.28).

Written in summation notation, Formula (12.28) would be

$$SSC = \left(\sum_{i=1}^{k}\frac{T_{i.}^2}{n_i}\right) - \frac{T_{..}^2}{n} \tag{12.29}$$

The within sum of squares is found by subtraction, as before:

$$SSE = SST - SSC$$

The degrees of freedom associated with the various sums of squares when the sample sizes are unequal are also analogous to those when the sample sizes are equal, just as the sums of squares in the two cases are analogous.

The degree of freedom associated with the total sum of squares is 1 less than the total number of observations, namely, $n - 1$.

The degrees of freedom associated with the among samples sum of squares is 1 less than the number of samples, namely, $k - 1$.

The number of degrees of freedom associated with the within samples sum of squares can be found as follows: the number of degrees if freedom within each sample is 1 less than the number of observations in that sample. Thus, there are $m_1 - 1$ degrees of freedom within Sample 1, $n_2 - 1$ degrees of freedom within Sample 2, and so on. The number of degrees of freedom within all the samples equals the sum of the degrees of freedom for each separate sample,

$$(m_1 - 1) + (n_2 - 1) + \ldots + (n_k - 1)$$

which equals

$$(m_1 + n_2 + \ldots + n_k) - k$$

which equals

$$n - k$$

The number of degrees of freedom associated with the within sum of squares can also be obtained by subtraction, because

$$\text{Total d.f.} = \text{Among d.f.} + \text{Within d.f.}$$

When we solve this relationship for Within d.f. we obtain

$$\text{Within d.f.} = \text{Total d.f.} - \text{Among d.f.} \qquad (12.30)$$

Although we have already obtained the Within d.f., we will obtain the Within d.f. again in order to illustrate the use of Formula (12.30). We need to use

$$\text{Total d.f.} = n - 1$$

and

$$\text{Among d.f.} = k - 1$$

when these values are substituted into Formula (12.30) we obtain

$$\text{Within d.f.} = (n - 1) - (k - 1)$$
$$= n - 1 - k + 1$$
$$= n - k$$

which agrees with the previously obtained value of Within d.f.

Table 12.5

Source	d.f.	Sum of squares	Mean squares
Among	$k - 1$	$SSC = \left(\sum_{i=1}^{k} \frac{T_{i\cdot}^2}{n_i} \right) - \frac{T_{\cdot\cdot}^2}{n}$	$MSC = \dfrac{SSC}{k-1}$
Error	$n - k$	$SSE = SST - SSC$	$MSE = \dfrac{SSE}{n-k}$
Total	$n - 1$	$SST = \left(\sum \sum x_{ij}^2 \right) - \frac{T_{\cdot\cdot}^2}{n}$	

The general analysis of variance table for a one-way analysis of variance with unequal sample sizes is shown in Table 12.5.

Procedure: One-way Analysis of Variance, Unequal Sample Sizes

    (1) Formulate the null and alternative hypotheses:

$$H_0: \mu_1 = \mu_2 = \ldots = \mu_k$$

and $\qquad\qquad H_1$: the means are not all equal

    (2) Decide upon an α-level.
        Find $F_\alpha(k - 1, n - k)$.

    (3) Obtain a random sample of size n_1 from Population 1, a random sample of size n_2 from Population 2, and so on to Population k, from which a sample of size n_k is obtained. (Or, equivalently, apply Treatment 1 to n_1 'members', Treatment 2 to n_2 'members', and so on.)

    (4) Calculate $T_1., T_2., \ldots, T_k.$
        Calculate $T..$
        Calculate $\dfrac{T_{..}^2}{n}$
        Calculate $\sum\sum x_{ij}^2$

    (5) Calculate $SST = \left(\sum\sum x_{ij}^2\right) - \dfrac{T_{..}^2}{n}$

        Calculate $SSC = \left(\dfrac{T_{1.}^2}{n_1} + \dfrac{T_{2.}^2}{n_2} + \ldots + \dfrac{T_{k.}^2}{n_k}\right) - \dfrac{T_{..}^2}{n}$

        Calculate $SSE = SST - SSC$.

    (6) Complete the analysis of variance table.

    (7) Calculate $F = \dfrac{MSC}{MSE}$

    (8) Reject H_0 if $F \geqslant F_\alpha(k - 1, n - k)$.

EXAMPLE: Five different treatments of fertilizer were applied to a number of plots of corn. Treatment 1 was applied to four plots, Treatments 2 and 3 were applied to six plots, Treatment 4 was applied to seven plots, and Treatment 5 was applied to three plots. The yields per plot, in bushels per are, are shown in Table 12.6. Is there any difference in the mean yield for the various treatments?

Table 12.6

(1)	(2)	(3)	(4)	(5)
78·9	63·5	79·1	87·0	75·9
72·3	74·1	90·3	91·2	77·2
81·1	75·5	85·6	75·3	81·5
85·7	80·8	81·4	79·4	
	71·3	74·5	80·7	
	79·4	95·3	82·8	
			89·6	

    (1) We are testing $\qquad H_0: \mu_1 = \mu_2 = \mu_3 = \mu_4 = \mu_5$
against $\qquad\qquad\qquad H_1$: the means are not all equal

    (2) We will use $\alpha = 0\cdot01$.

$$F_\alpha(k - 1, n - k) = F_{0.01}(4, 21) = 4\cdot37$$

(3) The observations resulting from the application of the five treatments are considered as random samples from the population of all possible yields.

(4)
$$T_1. = 318 \cdot 0$$
$$T_2. = 444 \cdot 6$$
$$T_3. = 506 \cdot 2$$
$$T_4. = 586 \cdot 0$$
$$T_5. = 234 \cdot 6$$
$$T.. = 2089 \cdot 4$$
$$\frac{T..^2}{n} = \frac{(2089 \cdot 4)^2}{26} = \frac{4\,365\,592 \cdot 36}{26} = 167\,907 \cdot 40$$

$$\sum_{i=1}^{k} \sum_{j=1}^{n} x_{ij}^2 = (78 \cdot 9)^2 + (72 \cdot 3)^2 + \ldots + (77 \cdot 2)^2 + (81 \cdot 5)^2$$
$$= 169\,131 \cdot 04$$

(5) $SST = 169\,131 \cdot 04 - 167\,907 \cdot 40 = 1223 \cdot 64$

$$SSC = \frac{(318 \cdot 0)^2}{4} + \frac{(444 \cdot 6)^2}{6} + \frac{(506 \cdot 2)^2}{6} + \frac{(586 \cdot 0)^2}{7} + \frac{(234 \cdot 6)^2}{3} - 167\,907 \cdot 40$$
$$= 25\,281 \cdot 00 + 32\,944 \cdot 86 + 42\,706 \cdot 41 + 49\,056 \cdot 57 + 18\,345 \cdot 72$$
$$\qquad\qquad - 167\,907 \cdot 40$$
$$= 168\,334 \cdot 56 - 167\,907 \cdot 40$$
$$= 427 \cdot 16$$
$$SSE = 1223 \cdot 64 - 427 \cdot 16 = 796 \cdot 48$$

(6) Analysis of variance table:

Source	d.f.	Sum of squares	Mean square
Among	4	427·16	106·79
Within	21	796·48	37·93
Total	25	1223·64	—

(7)
$$F = \frac{106 \cdot 79}{37 \cdot 93} = 2 \cdot 82$$

(8) $2 \cdot 82 < 4 \cdot 37$. We accept H_o.

TWO-WAY ANALYSIS OF VARIANCE

Sometimes an experimenter will be interested in two kinds of treatments and will want to test, simultaneously, whether there are differences between the means of Treatment 1 and whether there are differences between the means of Treatment 2.

For example, suppose that a department of education is interested in finding the best way to teach a certain topic in elementary mathematics. There are three methods of teaching the topic and four different ways in which to arrange the work pattern. Suppose that the same teacher, who is equally expert in all three methods, instructs each of twelve students of equal ability in this topic. Each student is taught by means of one of the three methods, and the work is done under one of the four work patterns.

At the end of the time period, suppose that the students were given a test, and the scores were those recorded in Table 12.7.

Table 12.7

		Method		
		1	2	3
Work pattern	1	58	56	65
	2	49	54	52
	3	60	71	39
	4	76	58	49

The total variation present can be partitioned into three independent pieces: an among row means sum of squares, an among column means sum of squares, and a within sum of squares. The among column means sum of squares, or among columns sum of squares, is found as before. The among rows sum of squares is an unfamiliar quantity, but it is found exactly as the among columns sum of squares is found. The total sum of squares is found exactly as it has been found in the two previous sections. The within mean square is found by subtraction, just as it was earlier.

In general, there will be k levels of Treatment 1 (these are the k columns), and r levels of Treatment 2 (these are the r rows). In the teaching illustration Treatment 1 refers to the different methods. There are three levels of Treatment 1 in this illustration, and any student who has been taught by Method 3 has received the third level of Treatment 1. Treatment 2 refers to the four different work patterns; there are four different levels of Treatment 2. Any student who has followed the first work pattern has received Level 1 of Treatment 2.

The general display of experimental results will resemble Table 12.7. We need row totals as well as column totals. We use the same notation as we did earlier. We replace the subscript over which we are summing by a dot.

Table 12.8

		Treatment 1			Totals
		1	2 ..... k		Totals
Treatment 2	1	x_{11}	$x_{21} .... x_{k1}$		$T_{.1}$
	2	x_{12}	$x_{22} .... x_{k2}$		$T_{.2}$
	.	.	.	.	.
	.	.	.	.	.
	.	.	.	.	.
	r	x_{1r}	$x_{2r} .... x_{kr}$		$T_{.r}$
Totals		$T_{1.}$	$T_{2.} .... T_{k.}$		$T_{..}$

The total of the first row is the sum

$$x_{11} + x_{21} + ... + x_{k1}$$

which can be written as $\sum_{i=1}^{k} x_{i1}$

since we are summing over the first subscript, our notation for this sum is $T._1$,

$$T._1 = \sum_{i=1}^{k} x_{i1}$$

Similarly, the sum of the second row is

$$T._2 = \sum_{i=1}^{k} x_{i2}$$

and the sum of the rth row is $\quad T._r = \sum_{i=1}^{k} x_{ir}$

In general, the total of the jth row is denoted $T._j$.

The among columns sum of squares, denoted SSC, is given by the same formula as before

$$SSC = \frac{T_1^2.}{r} + \frac{T_2^2.}{r} + \ldots + \frac{T_k^2.}{r} - \frac{T_{..}^2}{rk} \qquad (12.31)$$

The computing formula for the among rows sum of squares, denoted SSR, is analogous to Formula (12.31): the square of each row total is divided by the number of observations summed to obtain that total; these quotients are summed; and $T_{..}^2/rk$ is subtracted from the sum. The computing formula is

$$SSR = \frac{T._1^2}{k} + \frac{T._2^2}{k} + \ldots + \frac{T._r^2}{k} - \frac{T_{..}^2}{rk} \qquad (12.32)$$

The formula for the total sum of squares is

$$SST = \left(\sum \sum x_{ij}^2\right) - \frac{T_{..}^2}{rk} \qquad (12.33)$$

The formula for the within mean square is

$$SSE = SST - SSR - SSC \qquad (12.34)$$

The analysis of variance table for a two-way analysis of variance with one observation for each combination of levels of the two treatments is shown in Table 12.9.

Table 12.9

Source	d.f.	Sum of squares	Mean square
Among rows	$r-1$	$SSR = \left(\sum_{j=1}^{r} \frac{T._j^2}{k}\right) - \frac{T_{..}^2}{rk}$	$MSR = \dfrac{SSR}{r-1}$
Among columns	$k-1$	$SSC = \left(\sum_{i=1}^{k} \frac{T_{i.}^2}{r}\right) - \frac{T_{..}^2}{rk}$	$MSC = \dfrac{SSC}{k-1}$
Within	$(r-1)(k-1)$	$SSE = SST - SSR - SSC$	$MSE = \dfrac{SSE}{(r-1)(k-1)}$
Total	$rk-1$	$SST = \sum \sum x_{ij}^2 - \dfrac{T_{..}^2}{rk}$	

Procedure: Two-way Analysis of Variance

(1) Formulate the null and alternative hypotheses:

H_0: the row means are equal

against H_1: the row means are not equal

and H_0: the column means are equal

against H_1: the column means are not equal

Both of these null hypotheses can be tested by means of one analysis of variance. Decide upon a level of significance, α.

(2) There are k levels of Treatment 1 and r levels of Treatment 2. Each of these rk combinations is applied to one of rk members (plots of beans, or items produced in a factory, or persons, etc.), yielding an array of rk observations similar to that in Table 12.8.

(3) Calculate $T_{.1}, T_{.2}, \ldots,$ and $T_{.r}$

Calculate $\dfrac{T_{.1}^2}{k} + \dfrac{T_{.2}^2}{k} + \ldots + \dfrac{T_{.r}^2}{k}$

Calculate $T_{..} = T_{.1} + T_{.2} + \ldots + T_{.r}$

Calculate $\dfrac{T_{..}^2}{rk}$

Calculate $SSR = \left(\dfrac{T_{.1}^2}{k} + \dfrac{T_{.2}^2}{k} + \ldots + \dfrac{T_{.r}^2}{k} \right) - \dfrac{T_{..}^2}{rk}$

(4) Calculate $T_{1.}, T_{2.}, \ldots, T_{k.}$.

Calculate $\dfrac{T_{1.}^2}{r} + \dfrac{T_{2.}^2}{r} + \ldots + \dfrac{T_{k.}^2}{r}$

Calculate $SSC = \left(\dfrac{T_{1.}^2}{r} + \dfrac{T_{2.}^2}{r} + \ldots + \dfrac{T_{k.}^2}{r} \right) - \dfrac{T_{..}^2}{rk}$

(5) Calculate $\sum \sum x_{ij}^2$

Calculate $SST = (\sum \sum x_{ij}^2) - \dfrac{T_{..}^2}{rk}$

(6) Calculate $SSE = SST - SSR - SSC$.

(7) Complete the analysis of variance tables.

(8) Find $F_\alpha[r - 1, (r - 1)(k - 1)]$ in the F-table.

Compute $F = \dfrac{MSR}{MSE}$

Reject H_0: the row means are equal, if $F \geqslant F_\alpha[r - 1, (r - 1)(k - 1)]$.

(9) Find $F_\alpha[k - 1, (r - 1)(k - 1)]$ in the F-table.

Compute $F = \dfrac{MSC}{MSE}$

Reject H_0: the column means are equal, if

$$F \geqslant F_\alpha[k - 1, (r - 1)(k - 1)]$$

EXAMPLE: Use the data in Table 12·7 to determine whether there is any difference among teaching methods, and whether there is any difference among work patterns.

(1) We are to test:

H_0: the row means are equal

against H_1: the row means are not equal

and H_0: the column means are equal

against H_1: the column means are not equal

Let $\alpha = 0.05$.

(2) The experiment has already been described, and the results are displayed in Table 12.7.

(3)
$$T_{.1} = 179$$
$$T_{.2} = 155$$
$$T_{.3} = 170$$
$$T_{.4} = 183$$
$$T_{..} = 179 + 155 + 170 + 183 = 687$$

$$\frac{T_{.1}^2}{k} + \frac{T_{.2}^2}{k} + \frac{T_{.3}^2}{k} + \frac{T_{.4}^2}{k} = \frac{(179)^2}{3} + \frac{(155)^2}{3} + \frac{(170)^2}{3} + \frac{(183)^2}{3}$$

$$= 10\,680 \cdot 33 + 8008 \cdot 33 + 9633 \cdot 33 + 11\,163 \cdot 00$$

$$= 39\,484 \cdot 99$$

$$\frac{T_{..}^2}{rk} = \frac{(687)^2}{12} = 39\,330 \cdot 75$$

$$SSR = 39\,484 \cdot 99 - 39\,330 \cdot 75 = 154 \cdot 24$$

(4)
$$T_{1.} = 243$$
$$T_{2.} = 239$$
$$T_{3.} = 205$$

$$\frac{T_{1.}^2}{r} + \frac{T_{2.}^2}{r} + \frac{T_{3.}^2}{r} = \frac{(243)^2}{4} + \frac{(239)^2}{4} + \frac{(205)^2}{4}$$

$$= 14\,762 \cdot 25 + 14\,280 \cdot 25 + 10\,506 \cdot 25$$

$$= 39\,548 \cdot 75$$

$$SSC = 39\,548 \cdot 75 - 39\,330 \cdot 75 = 218 \cdot 00$$

(5)
$$\sum\sum x_{ij}^2 = 40\,449$$

$$SST = 40\,449 \cdot 00 - 39\,330 \cdot 75 = 1\,118 \cdot 25$$

(6)
$$SSE = 1118 \cdot 25 - 154 \cdot 24 - 218 \cdot 00 = 746 \cdot 01$$

(7) Analysis of variance table:

Source	d.f.	Sum of squares	Mean square
Among rows	3	154·24	51·41
Among columns	2	218·00	109·00
Within	6	746·01	124·34
Total	11	1118·25	

(8)
$$F_\alpha[r - 1, (r - 1)(k - 1)] = F_{0.05}(3, 6) = 4 \cdot 76$$

$$F = \frac{MSR}{MSE} = \frac{51 \cdot 41}{124 \cdot 34} = 0 \cdot 41$$

$F < F_{0.05}(3, 6)$. Accept H_o: row means are equal

(9)
$$F_\alpha[k - 1, (r - 1)(k - 1)] = F_{0.05}(2, 6) = 5 \cdot 14$$

$$F = \frac{MSC}{MSE} = \frac{109 \cdot 00}{124 \cdot 34} = 0 \cdot 88$$

$F < F_{0.05}(2, 6)$. Accept H_o: the column means are equal

EXERCISES

1. The three samples below have been obtained from normal populations with equal variances. Test the hypothesis that the population means are equal.

8	7	12
10	5	9
7	10	13
14	9	12
11	9	14

2. The four samples below have been obtained from normal populations with equal variances. Test the hypothesis that the population means are equal.

15	12	14	10
17	10	9	14
14	13	7	13
11	17	10	15
	14	8	12
		7	

3. For the data below, test that the column means are equal and that the row means are equal. Let $\alpha = 0.01$.

		Treatment 1			
		(1)	(2)	(3)	(4)
	(1)	58	65	53	68
Treatment 2	(2)	71	62	68	75
	(3)	66	54	60	67
	(4)	53	59	67	61

4. Perform a two-way analysis of variance on the data below.

		Treatment 1		
		(1)	(2)	(3)
	(1)	30	26	38
	(2)	24	29	28
Treatment 2	(3)	33	24	35
	(4)	36	31	30
	(5)	27	35	33

APPENDIX

LIST OF SELECTED SYMBOLS

Symbol	Page	Symbol	Page
$A(z)$	84	π	65
α	102	$\Phi(z)$	85
$b(x; n, \pi)$	66	r	132, 178
β	102	ρ	132
c	20, 178	R	190
$_nC_r$	60	r_s	180
$C(n, r)$	60	s	31
χ^2	168, 171	S	202
$\chi_\alpha^2(n-1)$	173	s^2	31, 75
$\bar{E}$	53	s_p^2	157, 196
$E(\mathbf{x})$	74	SSC	204
f_i	10	SSE	204
F	205	SSR	214
$F\alpha(n_1 - 1, n_2 - 1)$	199	SST	202
H_0	102	σ	75, 89
H_1	102	σ^2	75
k	174, 197	$\sum$	22
M	19	t	114
M.A.D.	29	T	187
MSC	205	$t_\alpha(n-1)$	115
MSE	205	$T_i.$	200
MSR	214	$T._j$	213, 214
μ	73	$T..$	202
n	17	$\mathbf{x}$	83
$n!$	57	x_i	17
$N(\mu, \sigma^2)$	89, 109	$\bar{x}$	17
C_r^n	60	x_i'	25
$\binom{n}{r}$	60, 92	$\hat{y}$	142
P_r^n	57	$\mathbf{z}$	83
$\mathbf{p}$	156	$z_{\alpha/2}$	104
$\hat{p}$	157	z_f	135
$P(\ )$	43	$\geqslant$	97
$P(B\|A)$	45	$\| \ \|$	29
$P(AB)$	45	$\leqslant$	97, 102
$P(A + B)$	48	$\cong$	156
		$\neq$	95, 156
			103

218

t-DISTRIBUTION

Degrees of freedom	$\alpha = 0.10$	$\alpha = 0.05$	$\alpha = 0.025$	$\alpha = 0.01$	$\alpha = 0.005$
1	3·078	6·314	12·706	31·821	63·657
2	1·886	2·920	4·303	6·965	9·925
3	1·638	2·353	3·182	4·541	5·841
4	1·533	2·132	2·776	3·747	4·604
5	1·476	2·015	2·571	3·365	4·032
6	1·440	1·943	2·447	3·143	3·707
7	1·415	1·895	2·365	2·998	3·499
8	1·397	1·860	2·306	2·896	3·355
9	1·383	1·833	2·262	2·821	3·250
10	1·372	1·812	2·228	2·764	3·169
11	1·363	1·796	2·201	2·718	3·106
12	1·356	1·782	2·179	2·681	3·055
13	1·350	1·771	2·160	2·650	3·012
14	1·345	1·761	2·145	2·624	2·977
15	1·341	1·753	2·131	2·602	2·947
16	1·337	1·746	2·120	2·583	2·921
17	1·333	1·740	2·110	2·567	2·898
18	1·330	1·734	2·101	2·552	2·878
19	1·328	1·729	2·093	2·539	2·861
20	1·325	1·725	2·086	2·528	2·845
21	1·323	1·721	2·080	2·518	2·831
22	1·321	1·717	2·074	2·508	2·819
23	1·319	1·714	2·069	2·500	2·807
24	1·318	1·711	2·064	2·492	2·797
25	1·316	1·708	2·060	2·485	2·787
26	1·315	1·706	2·056	2·479	2·779
27	1·314	1·703	2·052	2·473	2·771
28	1·313	1·701	2·048	2·467	2·763
29	1·311	1·699	2·045	2·462	2·756
30	1·310	1·697	2·042	2·457	2·750
40	1·303	1·684	2·021	2·423	2·704
60	1·296	1·671	2·000	2·390	2·660
120	1·289	1·658	1·980	2·358	2·617
∞	1·282	1·645	1·960	2·326	2·576

	0	1	2	3	4	5	6	7	8	9	1	2	3	4	5	6	7	8
10	1000	1020	1040	1061	1082	1103	1124	1145	1166	1188	2	4	6	8	10	13	15	17
11	1210	1232	1254	1277	1300	1323	1346	1369	1392	1416	2	5	7	9	11	14	16	18
12	1440	1464	1488	1513	1538	1563	1588	1613	1638	1664	2	5	7	10	12	15	17	20
13	1690	1716	1742	1769	1796	1823	1850	1877	1904	1932	3	5	8	11	13	16	19	22
14	1960	1988	2016	2045	2074	2103	2132	2161	2190	2220	3	6	9	12	14	17	20	23
15	2250	2280	2310	2341	2372	2403	2434	2465	2496	2528	3	6	9	12	15	19	22	25
16	2560	2592	2624	2657	2690	2723	2756	2789	2822	2856	3	7	10	13	16	20	23	26
17	2890	2924	2958	2993	3028	3063	3098	3133	3168	3204	3	7	10	14	17	21	24	28
18	3240	3276	3312	3349	3386	3423	3460	3497	3534	3572	4	7	11	15	18	22	26	30
19	3610	3648	3686	3725	3764	3803	3842	3881	3920	3960	4	8	12	16	19	23	27	31
20	4000	4040	4080	4121	4162	4203	4244	4285	4326	4368	4	8	12	16	20	25	29	33
21	4410	4452	4494	4537	4580	4623	4666	4709	4752	4796	4	9	13	17	21	26	30	34
22	4840	4884	4928	4973	5018	5063	5108	5153	5198	5244	4	9	13	18	22	27	31	36
23	5290	5336	5382	5429	5476	5523	5570	5617	5664	5712	5	9	14	19	23	28	33	38
24	5760	5808	5856	5905	5954	6003	6052	6101	6150	6200	5	10	15	20	24	29	34	39
25	6250	6300	6350	6401	6452	6503	6554	6605	6656	6708	5	10	15	20	25	31	36	41
26	6760	6812	6864	6917	6970	7023	7076	7129	7182	7236	5	11	16	21	26	32	37	42
27	7290	7344	7398	7453	7508	7563	7618	7673	7728	7784	5	11	16	22	28	33	38	44
28	7840	7896	7952	8009	8066	8123	8180	8237	8294	8352	6	11	17	23	28	34	40	46
29	8410	8468	8526	8585	8644	8703	8762	8821	8880	8940	6	12	18	24	30	35	41	47
30	9000	9060	9120	9181	9242	9303	9364	9425	9486	9548	6	12	18	24	31	37	43	49
31	9610	9672	9734	9797	9860	9923	9986				6	13	19	25	31	38	44	50
31								1005	1011	1018	1	1	2	3	3	4	5	5
32	1024	1030	1037	1043	1050	1056	1063	1069	1076	1082	1	1	2	3	3	4	5	5
33	1089	1096	1102	1109	1116	1122	1129	1136	1142	1149	1	1	2	3	3	4	5	5
34	1156	1163	1170	1176	1183	1190	1197	1204	1211	1218	1	1	2	3	3	4	5	6
35	1225	1232	1239	1246	1253	1260	1267	1274	1282	1289	1	1	2	3	4	4	5	6
36	1296	1303	1310	1318	1325	1332	1340	1347	1354	1362	1	1	2	3	4	4	5	6
37	1369	1376	1384	1391	1399	1406	1414	1421	1429	1436	1	2	2	3	4	5	5	6
38	1444	1452	1459	1467	1475	1482	1490	1498	1505	1513	1	2	2	3	4	5	6	6
39	1521	1529	1537	1544	1552	1560	1568	1576	1584	1592	1	2	2	3	4	5	6	6
40	1600	1608	1616	1624	1632	1640	1648	1656	1665	1673	1	2	2	3	4	5	6	6
41	1681	1689	1697	1706	1714	1722	1731	1739	1747	1756	1	2	2	3	4	5	6	7
42	1764	1772	1781	1789	1798	1806	1815	1823	1832	1840	1	2	3	3	4	5	6	7
43	1849	1858	1866	1875	1884	1892	1901	1910	1918	1927	1	2	3	3	4	5	6	7
44	1936	1945	1954	1962	1971	1980	1989	1998	2007	2016	1	2	3	4	5	5	6	7
45	2025	2034	2043	2052	2061	2070	2079	2088	2098	2107	1	2	3	4	5	5	6	7
46	2116	2125	2134	2144	2153	2162	2172	2181	2190	2200	1	2	3	4	5	6	7	7
47	2209	2218	2228	2237	2247	2256	2266	2275	2285	2294	1	2	3	4	5	6	7	8
48	2304	2314	2323	2333	2343	2352	2362	2372	2381	2391	1	2	3	4	5	6	7	8
49	2401	2411	2421	2430	2440	2450	2460	2470	2480	2490	1	2	3	4	5	6	7	8
50	2500	2510	2520	2530	2540	2550	2560	2570	2581	2591	1	2	3	4	5	6	7	8
51	2601	2611	2621	2632	2642	2652	2663	2673	2683	2694	1	2	3	4	5	6	7	8
52	2704	2714	2725	2735	2746	2756	2767	2777	2788	2798	1	2	3	4	5	6	7	8
53	2809	2820	2830	2841	2852	2862	2873	2884	2894	2905	1	2	3	4	5	6	7	9
54	2916	2927	2938	2948	2959	2970	2981	2992	3003	3014	1	2	3	4	6	7	8	9
	0	1	2	3	4	5	6	7	8	9	1	2	3	4	5	6	7	8

The position of the decimal point must be determined by inspection.

	0	1	2	3	4	5	6	7	8	9	1	2	3	4	5	6	7	8	9
5	3025	3036	3047	3058	3069	3080	3091	3102	3114	3125	1	2	3	4	6	7	8	9	10
6	3136	3147	3158	3170	3181	3192	3204	3215	3226	3238	1	2	3	5	6	7	8	9	10
7	3249	3260	3272	3283	3295	3306	3318	3329	3341	3352	1	2	3	5	6	7	8	9	10
8	3364	3376	3387	3399	3411	3422	3434	3446	3457	3469	1	2	4	5	6	7	8	9	11
9	3481	3493	3505	3516	3528	3540	3552	3564	3576	3588	1	2	4	5	6	7	8	10	11
0	3600	3612	3624	3636	3648	3660	3672	3684	3697	3709	1	2	4	5	6	7	8	10	11
1	3721	3733	3745	3758	3770	3782	3795	3807	3819	3832	1	2	4	5	6	7	9	10	11
2	3844	3856	3869	3881	3894	3906	3919	3931	3944	3956	1	3	4	5	6	8	9	10	11
3	3969	3982	3994	4007	4020	4032	4045	4058	4070	4083	1	3	4	5	6	8	9	10	11
4	4096	4109	4122	4134	4147	4160	4173	4186	4199	4212	1	3	4	5	6	8	9	10	12
5	4225	4238	4251	4264	4277	4290	4303	4316	4330	4343	1	3	4	5	7	8	9	10	12
6	4356	4369	4382	4396	4409	4422	4436	4449	4462	4476	1	3	4	5	7	8	9	11	12
7	4489	4502	4516	4529	4543	4556	4570	4583	4597	4610	1	3	4	5	7	8	9	11	12
8	4624	4638	4651	4665	4679	4692	4706	4720	4733	4747	1	3	4	5	7	8	10	11	12
9	4761	4775	4789	4802	4816	4830	4844	4858	4872	4886	1	3	4	6	7	8	10	11	13
0	4900	4914	4928	4942	4956	4970	4984	4998	5013	5027	1	3	4	6	7	8	10	11	13
1	5041	5055	5069	5084	5098	5112	5127	5141	5155	5170	1	3	4	6	7	9	10	11	13
2	5184	5198	5213	5227	5242	5256	5271	5285	5300	5314	1	3	4	6	7	9	10	11	13
3	5329	5344	5358	5373	5388	5402	5417	5432	5446	5461	1	3	4	6	7	9	10	12	13
4	5476	5491	5506	5520	5535	5550	5565	5580	5595	5610	1	3	4	6	7	9	10	12	13
5	5625	5640	5655	5670	5685	5700	5715	5730	5746	5761	2	3	5	6	8	9	11	12	14
6	5776	5791	5806	5822	5837	5852	5868	5883	5898	5914	2	3	5	6	8	9	11	12	14
7	5929	5944	5960	5975	5991	6006	6022	6037	6053	6068	2	3	5	6	8	9	11	12	14
8	6084	6100	6115	6131	6147	6162	6178	6194	6209	6225	2	3	5	6	8	9	11	13	14
9	6241	6257	6273	6288	6304	6320	6336	6352	6368	6384	2	3	5	6	8	10	11	13	14
0	6400	6416	6432	6448	6464	6480	6496	6512	6529	6545	2	3	5	6	8	10	11	13	14
1	6561	6577	6593	6610	6626	6642	6659	6675	6691	6708	2	3	5	7	8	10	11	13	15
2	6724	6740	6757	6773	6790	6806	6823	6839	6856	6872	2	3	5	7	8	10	12	13	15
3	6889	6906	6922	6939	6956	6972	6989	7006	7022	7039	2	3	5	7	8	10	12	13	15
4	7056	7073	7090	7106	7123	7140	7157	7174	7191	7208	2	3	5	7	8	10	12	14	15
5	7225	7242	7259	7276	7293	7310	7327	7344	7362	7379	2	3	5	7	9	10	12	14	15
6	7396	7413	7430	7448	7465	7482	7500	7517	7534	7552	2	3	5	7	9	10	12	14	16
7	7569	7586	7604	7621	7639	7656	7674	7691	7709	7726	2	4	5	7	9	11	12	14	16
8	7744	7762	7779	7797	7815	7832	7850	7868	7885	7903	2	4	5	7	9	11	12	14	16
9	7921	7939	7957	7974	7992	8010	8028	8046	8064	8082	2	4	5	7	9	11	13	14	16
0	8100	8118	8136	8154	8172	8190	8208	8226	8245	8263	2	4	5	7	9	11	13	14	16
1	8281	8299	8317	8336	8354	8372	8391	8409	8427	8446	2	4	5	7	9	11	13	15	16
2	8464	8482	8501	8519	8538	8556	8575	8593	8612	8630	2	4	6	7	9	11	13	15	17
3	8649	8668	8686	8705	8724	8742	8761	8780	8798	8817	2	4	6	7	9	11	13	15	17
4	8836	8855	8874	8892	8911	8930	8949	8968	8987	9006	2	4	6	8	9	11	13	15	17
5	9025	9044	9063	9082	9101	9120	9139	9158	9178	9197	2	4	6	8	10	11	13	15	17
6	9216	9235	9254	9274	9293	9312	9332	9351	9370	9390	2	4	6	8	10	12	14	15	17
7	9409	9428	9448	9467	9487	9506	9526	9545	9565	9584	2	4	6	8	10	12	14	16	18
8	9604	9624	9643	9663	9683	9702	9722	9742	9761	9781	2	4	6	8	10	12	14	16	18
9	9801	9821	9841	9860	9880	9900	9920	9940	9960	9980	2	4	6	8	10	12	14	16	18
	0	1	2	3	4	5	6	7	8	9	1	2	3	4	5	6	7	8	9

The position of the decimal point must be determined by inspection.

Square Roots

	0	1	2	3	4	5	6	7	8	9	1	2	3	4	5	6	7	8	9
10	1000	1005	1010	1015	1020	1025	1030	1034	1039	1044	0	1	1	2	2	3	3	4	4
	3162	3178	3194	3209	3225	3240	3256	3271	3286	3302	2	3	5	6	8	9	11	12	14
11	1049	1054	1058	1063	1068	1072	1077	1082	1086	1091	0	1	1	2	2	3	3	4	4
	3317	3332	3347	3362	3376	3391	3406	3421	3435	3450	1	3	4	6	7	9	10	12	13
12	1095	1100	1105	1109	1114	1118	1122	1127	1131	1136	0	1	1	2	2	3	3	4	4
	3464	3479	3493	3507	3521	3536	3550	3564	3578	3592	1	3	4	6	7	8	10	11	13
13	1140	1145	1149	1153	1158	1162	1166	1170	1175	1179	0	1	1	2	2	3	3	3	4
	3606	3619	3633	3647	3661	3674	3688	3701	3715	3728	1	3	4	5	7	8	10	11	12
14	1183	1187	1192	1196	1200	1204	1208	1212	1217	1221	0	1	1	2	2	3	3	3	4
	3742	3755	3768	3782	3795	3808	3821	3834	3847	3860	1	3	4	5	7	8	9	11	12
15	1225	1229	1233	1237	1241	1245	1249	1253	1257	1261	0	1	1	2	2	3	3	3	4
	3873	3886	3899	3912	3924	3937	3950	3962	3975	3987	1	3	4	5	6	8	9	10	11
16	1265	1269	1273	1277	1281	1285	1288	1292	1296	1300	0	1	1	2	2	3	3	3	4
	4000	4012	4025	4037	4050	4062	4074	4087	4099	4111	1	2	4	5	6	7	9	10	11
17	1304	1308	1311	1315	1319	1323	1327	1330	1334	1338	0	1	1	2	2	2	3	3	3
	4123	4135	4147	4159	4171	4183	4195	4207	4219	4231	1	2	4	5	6	7	8	10	11
18	1342	1345	1349	1353	1356	1360	1364	1367	1371	1375	0	1	1	1	2	2	3	3	4
	4243	4254	4266	4278	4290	4301	4313	4324	4336	4347	1	2	3	5	6	7	8	9	10
19	1378	1382	1386	1389	1393	1396	1400	1404	1407	1411	0	1	1	1	2	2	3	3	4
	4359	4370	4382	4393	4405	4416	4427	4438	4450	4461	1	2	3	5	6	7	8	9	10
20	1414	1418	1421	1425	1428	1432	1435	1439	1442	1446	0	1	1	1	2	2	2	3	3
	4472	4483	4494	4506	4517	4528	4539	4550	4561	4572	1	2	3	4	5	7	8	9	10
21	1449	1453	1456	1459	1463	1466	1470	1473	1476	1480	0	1	1	1	2	2	2	3	3
	4583	4593	4604	4615	4626	4637	4648	4658	4669	4680	1	2	3	4	5	6	8	9	10
22	1483	1487	1490	1493	1497	1500	1503	1507	1510	1513	0	1	1	1	2	2	2	3	3
	4690	4701	4712	4722	4733	4743	4754	4764	4775	4785	1	2	3	4	5	6	7	8	9
23	1517	1520	1523	1526	1530	1533	1536	1539	1543	1546	0	1	1	1	2	2	2	3	3
	4796	4806	4817	4827	4837	4848	4858	4868	4879	4889	1	2	3	4	5	6	7	8	9
24	1549	1552	1556	1559	1562	1565	1568	1572	1575	1578	0	1	1	1	2	2	2	3	3
	4899	4909	4919	4930	4940	4950	4960	4970	4980	4990	1	2	3	4	5	6	7	8	9
25	1581	1584	1587	1591	1594	1597	1600	1603	1606	1609	0	1	1	1	2	2	2	3	3
	5000	5010	5020	5030	5040	5050	5060	5070	5079	5089	1	2	3	4	5	6	7	8	9
26	1612	1616	1619	1622	1625	1628	1631	1634	1637	1640	0	1	1	1	2	2	2	2	3
	5099	5109	5119	5128	5138	5148	5158	5167	5177	5187	1	2	3	4	5	6	7	8	9
27	1643	1646	1649	1652	1655	1658	1661	1664	1667	1670	0	1	1	1	2	2	2	2	3
	5196	5206	5215	5225	5235	5244	5254	5263	5273	5282	1	2	3	4	5	6	7	8	9
28	1673	1676	1679	1682	1685	1688	1691	1694	1697	1700	0	1	1	1	1	2	2	2	3
	5292	5301	5310	5320	5329	5339	5348	5357	5367	5376	1	2	3	4	5	6	7	7	8
29	1703	1706	1709	1712	1715	1718	1720	1723	1726	1729	0	1	1	1	1	2	2	2	3
	5385	5394	5404	5413	5422	5431	5441	5450	5459	5468	1	2	3	4	5	5	6	7	8
30	1732	1735	1738	1741	1744	1746	1749	1752	1755	1758	0	1	1	1	1	2	2	2	3
	5477	5486	5495	5505	5514	5523	5532	5541	5550	5559	1	2	3	4	4	5	6	7	8
31	1761	1764	1766	1769	1772	1775	1778	1780	1783	1786	0	1	1	1	1	2	2	2	3
	5568	5577	5586	5595	5604	5612	5621	5630	5639	5648	1	2	3	3	4	5	6	7	8
32	1789	1792	1794	1797	1800	1803	1806	1808	1811	1814	0	1	1	1	1	2	2	2	2
	5657	5666	5675	5683	5692	5701	5710	5718	5727	5736	1	2	3	3	4	5	6	7	8
	0	**1**	**2**	**3**	**4**	**5**	**6**	**7**	**8**	**9**	**1**	**2**	**3**	**4**	**5**	**6**	**7**	**8**	**9**

The first significant figure and the position of the decimal point must
be determined by inspection.

	0	1	2	3	4	5	6	7	8	9	1	2	3	4	5	6	7	8	9
33	1817	1819	1822	1825	1828	1830	1833	1836	1838	1841	0	1	1	1	1	2	2	2	2
	5745	5753	5762	5771	5779	5788	5797	5805	5814	5822	1	2	3	3	4	5	6	7	8
34	1844	1847	1849	1852	1855	1857	1860	1863	1865	1868	0	1	1	1	1	2	2	2	2
	5831	5840	5848	5857	5865	5874	5882	5891	5899	5908	1	2	3	3	4	5	6	7	8
35	1871	1873	1876	1879	1881	1884	1887	1889	1892	1895	0	1	1	1	1	2	2	2	2
	5916	5925	5933	5941	5950	5958	5967	5975	5983	5992	1	2	2	3	4	5	6	7	8
36	1897	1900	1903	1905	1908	1910	1913	1916	1918	1921	0	1	1	1	1	2	2	2	2
	6000	6008	6017	6025	6033	6042	6050	6058	6066	6075	1	2	2	3	4	5	6	7	7
37	1924	1926	1929	1931	1934	1936	1939	1942	1944	1947	0	1	1	1	1	2	2	2	2
	6083	6091	6099	6107	6116	6124	6132	6140	6148	6156	1	2	2	3	4	5	6	7	7
38	1949	1952	1954	1957	1960	1962	1965	1967	1970	1972	0	1	1	1	1	2	2	2	2
	6164	6173	6181	6189	6197	6205	6213	6221	6229	6237	1	2	2	3	4	5	6	6	7
39	1975	1977	1980	1982	1985	1987	1990	1992	1995	1997	0	1	1	1	1	2	2	2	2
	6245	6253	6261	6269	6277	6285	6293	6301	6309	6317	1	2	2	3	4	5	6	6	7
40	2000	2002	2005	2007	2010	2012	2015	2017	2020	2022	0	0	1	1	1	1	2	2	2
	6325	6332	6340	6348	6356	6364	6372	6380	6387	6395	1	2	2	3	4	5	6	6	7
41	2025	2027	2030	2032	2035	2037	2040	2042	2045	2047	0	0	1	1	1	1	2	2	2
	6403	6411	6419	6427	6434	6442	6450	6458	6465	6473	1	2	2	3	4	5	5	6	7
42	2049	2052	2054	2057	2059	2062	2064	2066	2069	2071	0	0	1	1	1	1	2	2	2
	6481	6488	6496	6504	6512	6519	6527	6535	6542	6550	1	2	2	3	4	5	5	6	7
43	2074	2076	2078	2081	2083	2086	2088	2090	2093	2095	0	0	1	1	1	1	2	2	2
	6557	6565	6573	6580	6588	6595	6603	6611	6618	6626	1	2	2	3	4	5	5	6	7
44	2098	2100	2102	2105	2107	2110	2112	2114	2117	2119	0	0	1	1	1	1	2	2	2
	6633	6641	6648	6656	6663	6671	6678	6686	6693	6701	1	2	2	3	4	4	5	6	7
45	2121	2124	2126	2128	2131	2133	2135	2138	2140	2142	0	0	1	1	1	1	2	2	2
	6708	6716	6723	6731	6738	6745	6753	6760	6768	6775	1	1	2	3	4	4	5	6	7
46	2145	2147	2149	2152	2154	2156	2159	2161	2163	2166	0	0	1	1	1	1	2	2	2
	6782	6790	6797	6804	6812	6819	6826	6834	6841	6848	1	1	2	3	4	4	5	6	7
47	2168	2170	2173	2175	2177	2179	2182	2184	2186	2189	0	0	1	1	1	1	2	2	2
	6856	6863	6870	6877	6885	6892	6899	6907	6914	6921	1	1	2	3	4	4	5	6	7
48	2191	2193	2195	2198	2200	2202	2205	2207	2209	2211	0	0	1	1	1	1	2	2	2
	6928	6935	6943	6950	6957	6964	6971	6979	6986	6993	1	1	2	3	4	4	5	6	6
49	2214	2216	2218	2220	2223	2225	2227	2229	2232	2234	0	0	1	1	1	1	2	2	2
	7000	7007	7014	7021	7029	7036	7043	7050	7057	7064	1	1	2	3	4	4	5	6	6
50	2236	2238	2241	2243	2245	2247	2249	2252	2254	2256	0	0	1	1	1	1	2	2	2
	7071	7078	7085	7092	7099	7106	7113	7120	7127	7134	1	1	2	3	4	4	5	6	6
51	2258	2261	2263	2265	2267	2269	2272	2274	2276	2278	0	0	1	1	1	1	2	2	2
	7141	7148	7155	7162	7169	7176	7183	7190	7197	7204	1	1	2	3	4	4	5	6	6
52	2280	2283	2285	2287	2289	2291	2293	2296	2298	2300	0	0	1	1	1	1	2	2	2
	7211	7218	7225	7232	7239	7246	7253	7259	7266	7273	1	1	2	3	3	4	5	6	6
53	2302	2304	2307	2309	2311	2313	2315	2317	2319	2322	0	0	1	1	1	1	2	2	2
	7280	7287	7294	7301	7308	7314	7321	7328	7335	7342	1	1	2	3	3	4	5	5	6
54	2324	2326	2328	2330	2332	2335	2337	2339	2341	2343	0	0	1	1	1	1	1	2	2
	7348	7355	7362	7369	7376	7382	7389	7396	7403	7409	1	1	2	3	3	4	5	5	6
	0	1	2	3	4	5	6	7	8	9	1	2	3	4	5	6	7	8	9

The first significant figure and the position of the decimal point must be determined by inspection.

	0	1	2	3	4	5	6	7	8	9	1 2 3	4 5 6	7 8 9
55	2345	2347	2349	2352	2354	2356	2358	2360	2362	2364	0 0 1	1 1 1	1 2 2
	7416	7423	7430	7436	7443	7450	7457	7463	7470	7477	1 1 2	3 3 4	5 5 6
56	2366	2369	2371	2373	2375	2377	2379	2381	2383	2385	0 0 1	1 1 1	1 2 2
	7483	7490	7497	7503	7510	7517	7523	7530	7537	7543	1 1 2	3 3 4	5 5 6
57	2387	2390	2392	2394	2396	2398	2400	2402	2404	2406	0 0 1	1 1 1	1 2 2
	7550	7556	7563	7570	7576	7583	7589	7596	7603	7609	1 1 2	3 3 4	5 5 6
58	2408	2410	2412	2415	2417	2419	2421	2423	2425	2427	0 0 1	1 1 1	1 2 2
	7616	7622	7629	7635	7642	7649	7655	7662	7668	7675	1 1 2	3 3 4	5 5 6
59	2429	2431	2433	2435	2437	2439	2441	2443	2445	2447	0 0 1	1 1 1	1 2 2
	7681	7688	7694	7701	7707	7714	7720	7727	7733	7740	1 1 2	3 3 4	5 5 6
60	2449	2452	2454	2456	2458	2460	2462	2464	2466	2468	0 0 1	1 1 1	1 2 2
	7746	7752	7759	7765	7772	7778	7785	7791	7797	7804	1 1 2	3 3 4	4 5 6
61	2470	2472	2474	2476	2478	2480	2482	2484	2486	2488	0 0 1	1 1 1	1 2 2
	7810	7817	7823	7829	7836	7842	7849	7855	7861	7868	1 1 2	3 3 4	4 5 6
62	2490	2492	2494	2496	2498	2500	2502	2504	2506	2508	0 0 1	1 1 1	1 2 2
	7874	7880	7887	7893	7899	7906	7912	7918	7925	7931	1 1 2	3 3 4	4 5 6
63	2510	2512	2514	2516	2518	2520	2522	2524	2526	2528	0 0 1	1 1 1	1 2 2
	7937	7944	7950	7956	7962	7969	7975	7981	7987	7994	1 1 2	3 3 4	4 5 6
64	2530	2532	2534	2536	2538	2540	2542	2544	2546	2548	0 0 1	1 1 1	1 2 2
	8000	8006	8012	8019	8025	8031	8037	8044	8050	8056	1 1 2	2 3 4	4 5 6
65	2550	2551	2553	2555	2557	2559	2561	2563	2565	2567	0 0 1	1 1 1	1 2 2
	8062	8068	8075	8081	8087	8093	8099	8106	8112	8118	1 1 2	2 3 4	4 5 5
66	2569	2571	2573	2575	2577	2579	2581	2583	2585	2587	0 0 1	1 1 1	1 2 2
	8124	8130	8136	8142	8149	8155	8161	8167	8173	8179	1 1 2	2 3 4	4 5 5
67	2588	2590	2592	2594	2596	2598	2600	2602	2604	2606	0 0 1	1 1 1	1 2 2
	8185	8191	8198	8204	8210	8216	8222	8228	8234	8240	1 1 2	2 3 4	4 5 5
68	2608	2610	2612	2613	2615	2617	2619	2621	2623	2625	0 0 1	1 1 1	1 2 2
	8246	8252	8258	8264	8270	8276	8283	8289	8295	8301	1 1 2	2 3 4	4 5 5
69	2627	2629	2631	2632	2634	2636	2638	2640	2642	2644	0 0 1	1 1 1	1 2 2
	8307	8313	8319	8325	8331	8337	8343	8349	8355	8361	1 1 2	2 3 4	4 5 5
70	2646	2648	2650	2651	2653	2655	2657	2659	2661	2663	0 0 1	1 1 1	1 2 2
	8367	8373	8379	8385	8390	8396	8402	8408	8414	8420	1 1 2	2 3 4	4 5 5
71	2665	2666	2668	2670	2672	2674	2676	2678	2680	2681	0 0 1	1 1 1	1 1 2
	8426	8432	8438	8444	8450	8456	8462	8468	8473	8479	1 1 2	2 3 3	4 5 5
72	2683	2685	2687	2689	2691	2693	2694	2696	2698	2700	0 0 1	1 1 1	1 1 2
	8485	8491	8497	8503	8509	8515	8521	8526	8532	8538	1 1 2	2 3 3	4 5 5
73	2702	2704	2706	2707	2709	2711	2713	2715	2717	2718	0 0 1	1 1 1	1 1 2
	8544	8550	8556	8562	8567	8573	8579	8585	8591	8597	1 1 2	2 3 3	4 5 5
74	2720	2722	2724	2726	2728	2729	2731	2733	2735	2737	0 0 1	1 1 1	1 1 2
	8602	8608	8614	8620	8626	8631	8637	8643	8649	8654	1 1 2	2 3 3	4 5 5
75	2739	2740	2742	2744	2746	2748	2750	2751	2753	2755	0 0 1	1 1 1	1 1 2
	8660	8666	8672	8678	8683	8689	8695	8701	8706	8712	1 1 2	2 3 3	4 5 5
76	2757	2759	2760	2762	2764	2766	2768	2769	2771	2773	0 0 1	1 1 1	1 1 2
	8718	8724	8729	8735	8741	8746	8752	8758	8764	8769	1 1 2	2 3 3	4 5 5
77	2775	2777	2778	2780	2782	2784	2786	2787	2789	2791	0 0 1	1 1 1	1 1 2
	8775	8781	8786	8792	8798	8803	8809	8815	8820	8826	1 1 2	2 3 3	4 4 5
	0	1	2	3	4	5	6	7	8	9	1 2 3	4 5 6	7 8 9

The first significant figure and the position of the decimal point must be determined by inspection.

	0	1	2	3	4	5	6	7	8	9	1 2 3	4 5 6	7 8 9
78	2793	2795	2796	2798	2800	2802	2804	2805	2807	2809	0 0 1	1 1 1	1 1 2
	8832	8837	8843	8849	8854	8860	8866	8871	8877	8883	1 1 2	2 3 3	4 4 5
79	2811	2812	2814	2816	2818	2820	2821	2823	2825	2827	0 0 1	1 1 1	1 1 2
	8888	8894	8899	8905	8911	8916	8922	8927	8933	8939	1 1 2	2 3 3	4 4 5
80	2828	2830	2832	2834	2835	2837	2839	2841	2843	2844	0 0 1	1 1 1	1 1 2
	8944	8950	8955	8961	8967	8972	8978	8983	8989	8994	1 1 2	2 3 3	4 4 5
81	2846	2848	2850	2851	2853	2855	2857	2858	2860	2862	0 0 1	1 1 1	1 1 2
	9000	9006	9011	9017	9022	9028	9033	9039	9044	9050	1 1 2	2 3 3	4 4 5
82	2864	2865	2867	2869	2871	2872	2874	2876	2877	2879	0 0 1	1 1 1	1 1 2
	9055	9061	9066	9072	9077	9083	9088	9094	9099	9105	1 1 2	2 3 3	4 4 5
83	2881	2883	2884	2886	2888	2890	2891	2893	2895	2897	0 0 1	1 1 1	1 1 2
	9110	9116	9121	9127	9132	9138	9143	9149	9154	9160	1 1 2	2 3 3	4 4 5
84	2898	2900	2902	2903	2905	2907	2909	2910	2912	2914	0 0 1	1 1 1	1 1 2
	9165	9171	9176	9182	9187	9192	9198	9203	9209	9214	1 1 2	2 3 3	4 4 5
85	2915	2917	2919	2921	2922	2924	2926	2927	2929	2931	0 0 1	1 1 1	1 1 2
	9220	9225	9230	9236	9241	9247	9252	9257	9263	9268	1 1 2	2 3 3	4 4 5
86	2933	2934	2936	2938	2939	2941	2943	2944	2946	2948	0 0 1	1 1 1	1 1 2
	9274	9279	9284	9290	9295	9301	9306	9311	9317	9322	1 1 2	2 3 3	4 4 5
87	2950	2951	2953	2955	2956	2958	2960	2961	2963	2965	0 0 1	1 1 1	1 1 2
	9327	9333	9338	9343	9349	9354	9359	9365	9370	9375	1 1 2	2 3 3	4 4 5
88	2966	2968	2970	2972	2973	2975	2977	2978	2980	2982	0 0 1	1 1 1	1 1 2
	9381	9386	9391	9397	9402	9407	9413	9418	9423	9429	1 1 2	2 3 3	4 4 5
89	2983	2985	2987	2988	2990	2992	2993	2995	2997	2998	0 0 1	1 1 1	1 1 2
	9434	9439	9445	9450	9455	9460	9466	9471	9476	9482	1 1 2	2 3 3	4 4 5
90	3000	3002	3003	3005	3007	3008	3010	3012	3013	3015	0 0 0	1 1 1	1 1 1
	9487	9492	9497	9503	9508	9513	9518	9524	9529	9534	1 1 2	2 3 3	4 4 5
91	3017	3018	3020	3022	3023	3025	3027	3028	3030	3032	0 0 0	1 1 1	1 1 1
	9539	9545	9550	9555	9560	9566	9571	9576	9581	9586	1 1 2	2 3 3	4 4 5
92	3033	3035	3036	3038	3040	3041	3043	3045	3046	3048	0 0 0	1 1 1	1 1 1
	9592	9597	9602	9607	9612	9618	9623	9628	9633	9638	1 1 2	2 3 3	4 4 5
93	3050	3051	3053	3055	3056	3058	3059	3061	3063	3064	0 0 0	1 1 1	1 1 1
	9644	9649	9654	9659	9664	9670	9675	9680	9685	9690	1 1 2	2 3 3	4 4 5
94	3066	3068	3069	3071	3072	3074	3076	3077	3079	3081	0 0 0	1 1 1	1 1 1
	9695	9701	9706	9711	9716	9721	9726	9731	9737	9742	1 1 2	2 3 3	4 4 5
95	3082	3084	3085	3087	3089	3090	3092	3094	3095	3097	0 0 0	1 1 1	1 1 1
	9747	9752	9757	9762	9767	9772	9778	9783	9788	9793	1 1 2	2 3 3	4 4 5
96	3098	3100	3102	3103	3105	3106	3108	3110	3111	3113	0 0 0	1 1 1	1 1 1
	9798	9803	9808	9813	9818	9823	9829	9834	9839	9844	1 1 2	2 3 3	4 4 5
97	3114	3116	3118	3119	3121	3122	3124	3126	3127	3129	0 0 0	1 1 1	1 1 1
	9849	9854	9859	9864	9869	9874	9879	9884	9889	9894	1 1 2	2 3 3	4 4 5
98	3130	3132	3134	3135	3137	3138	3140	3142	3143	3145	0 0 0	1 1 1	1 1 1
	9899	9905	9910	9915	9920	9925	9930	9935	9940	9945	0 1 1	2 2 3	3 4 4
99	3146	3148	3150	3151	3153	3154	3156	3158	3159	3161	0 0 0	1 1 1	1 1 1
	9950	9955	9960	9965	9970	9975	9980	9985	9990	9995	0 1 1	2 2 3	3 4 4
	0	**1**	**2**	**3**	**4**	**5**	**6**	**7**	**8**	**9**	**1 2 3**	**4 5 6**	**7 8 9**

The first significant figure and the position of the decimal point must
be determined by inspection.

Statistics Made Simple

AREA OF THE STANDARD NORMAL DISTRIBUTION

z	A(z)	z	A(z)	z	A(z)	z	A(z)
0·00	0·00000	0·50	0·19146	1·00	0·34134	1·50	0·43319
0·01	0·00399	0·51	0·19497	1·01	0·34375	1·51	0·43448
0·02	0·00798	0·52	0·19847	1·02	0·34614	1·52	0·43574
0·03	0·01197	0·53	0·20194	1·03	0·34849	1·53	0·43699
0·04	0·01595	0·54	0·20540	1·04	0·35083	1·54	0·43822
0·05	0·01994	0·55	0·20884	1·05	0·35314	1·55	0·43943
0·06	0·02392	0·56	0·21226	1·06	0·35543	1·56	0·44062
0·07	0·02790	0·57	0·21566	1·07	0·35769	1·57	0·44179
0·08	0·03188	0·58	0·21904	1·08	0·35993	1·58	0·44295
0·09	0·03586	0·59	0·22240	1·09	0·36214	1·59	0·44408
0·10	0·03983	0·60	0·22575	1·10	0·36433	1·60	0·44520
0·11	0·04380	0·61	0·22907	1·11	0·36650	1·61	0·44630
0·12	0·04776	0·62	0·23237	1·12	0·36864	1·62	0·44738
0·13	0·05172	0·63	0·23565	1·13	0·37076	1·63	0·44845
0·14	0·05567	0·64	0·23891	1·14	0·37286	1·64	0·44950
0·15	0·05962	0·65	0·24215	1·15	0·37493	1·65	0·45053
0·16	0·06356	0·66	0·24537	1·16	0·37698	1·66	0·45154
0·17	0·06750	0·67	0·24857	1·17	0·37900	1·67	0·45254
0·18	0·07142	0·68	0·25175	1·18	0·38100	1·68	0·45352
0·19	0·07535	0·69	0·25490	1·19	0·38298	1·69	0·45449
0·20	0·07926	0·70	0·25804	1·20	0·38493	1·70	0·45543
0·21	0·08317	0·71	0·26115	1·21	0·38686	1·71	0·45637
0·22	0·08706	0·72	0·26424	1·22	0·38877	1·72	0·45728
0·23	0·09095	0·73	0·26730	1·23	0·39065	1·73	0·45818
0·24	0·09483	0·74	0·27035	1·24	0·39251	1·74	0·45907
0·25	0·09871	0·75	0·27337	1.25	0·39435	1·75	0·45994
0·26	0·10257	0·76	0·27637	1·26	0·39617	1·76	0·46080
0·27	0·10642	0·77	0·27935	1·27	0·39796	1·77	0·46164
0·28	0·11026	0·78	0·28230	1·28	0·39973	1·78	0·46246
0·29	0·11409	0·79	0·28524	1·29	0·40147	1·79	0·46327
0·30	0·11791	0·80	0·28814	1·30	0·40320	1·80	0·46407
0·31	0·12172	0·81	0·29103	1·31	0·40490	1·81	0·46485
0·32	0·12552	0·82	0·29389	1·32	0·40658	1·82	0·46562
0·33	0·12930	0·83	0·29673	1·33	0·40824	1·83	0·46638
0·34	0·13307	0·84	0·29955	1·34	0·40988	1·84	0·46712
0·35	0·13683	0·85	0·30234	1·35	0·41149	1·85	0·46784
0·36	0·14058	0·86	0·30511	1·36	0·41309	1·86	0·46856
0·37	0·14431	0·87	0·30785	1·37	0·41466	1·87	0·46926
0·38	0·14803	0·88	0·31057	1·38	0·41621	1·88	0·46995
0·39	0·15173	0·89	0·31327	1·39	0·41774	1·89	0·47062
0·40	0·15542	0·90	0·31594	1·40	0·41924	1·90	0·47128
0·41	0·15910	0·91	0·31859	1·41	0·42073	1·91	0·47193
0·42	0·16276	0·92	0·32121	1·42	0·42220	1·92	0·47257
0·43	0·16640	0·93	0·32381	1·43	0·42364	1·93	0·47320
0·44	0·17003	0·94	0·32639	1·44	0·42507	1·94	0·47381
0·45	0·17364	0·95	0·32894	1·45	0·42647	1·95	0·47441
0·46	0·17724	0·96	0·33147	1·46	0·42785	1·96	0·47500
0·47	0·18082	0·97	0·33398	1·47	0·42922	1·97	0·47558
0·48	0·18439	0·98	0·33646	1·48	0·43056	1·98	0·47615
0·49	0·18793	0·99	0·33891	1·49	0·43189	1·99	0·47670

AREA OF THE STANDARD NORMAL DISTRIBUTION

z	A(z)	z	A(z)	z	A(z)	z	A(z)
2·00	0·47725	2·50	0·49379	3·00	0·49865	3·50	0·49977
2·01	0·47778	2·51	0·49396	3·01	0·49869	3·51	0·49978
2·02	0·47831	2·52	0·49413	3·02	0·49874	3·52	0·49978
2·03	0·47882	2·53	0·49430	3·03	0·49878	3·53	0·49979
2·04	0·47932	2·54	0·49446	3·04	0·49882	3·54	0·49980
2·05	0·47982	2·55	0·49461	3·05	0·49886	3·55	0·49981
2·06	0·48030	2·56	0·49477	3·06	0·49889	3·56	0·49981
2·07	0·48077	2·57	0·49492	3·07	0·49893	3·57	0·49982
2·08	0·48124	2·58	0·49506	3·08	0·49896	3·58	0·49983
2·09	0·48169	2·59	0·49520	3·09	0·49900	3·59	0·49983
2·10	0·48214	2·60	0·49534	3·10	0·49903	3·60	0·49984
2·11	0·48257	2·61	0·49547	3·11	0·49906	3·61	0·49985
2·12	0·48300	2·62	0·49560	3·12	0·49910	3·62	0·49985
2·13	0·48341	2·63	0·49573	3·13	0·49913	3.63	0·49986
2·14	0·48382	2·64	0·49585	3·14	0·49916	3·64	0·49986
2·15	0·48422	2·65	0·49598	3·15	0·49918	3·65	0·49987
2·16	0·48461	2·66	0·49609	3·16	0·49921	3·66	0·49987
2·17	0·48500	2·67	0·49621	3·17	0·49924	3·67	0·49988
2·18	0·48537	2·68	0·49632	3·18	0·49926	3·68	0·49988
2·19	0·48574	2·69	0·49643	3·19	0·49929	3·69	0·49989
2·20	0·48610	2·70	0·49653	3·20	0·49931	3·70	0·49989
2·21	0·48645	2·71	0·49664	3·21	0·49934	3·71	0·49990
2·22	0·48679	2·72	0·49674	3·22	0·49936	3·72	0·49990
2·23	0·48713	2·73	0·49683	3·23	0·49938	3·73	0·49990
2·24	0·48745	2·74	0·49693	3·24	0·49940	3·74	0·49991
2·25	0·48778	2·75	0·49702	3·25	0·49942	3·75	0·49991
2·26	0·48809	2·76	0·49711	3·26	0·49944	3·76	0·49992
2·27	0·48840	2·77	0·49720	3·27	0·49946	3·77	0·49992
2·28	0·48870	2·78	0·49728	3·28	0·49948	3·78	0·49992
2·29	0·48899	2·79	0·49736	3·29	0·49950	3·79	0·49992
2·30	0·48928	2·80	0·49744	3·30	0·49952	3·80	0·49993
2·31	0·48956	2·81	0·49752	3·31	0·49953	3·81	0·49993
2·32	0·48983	2·82	0·49760	3·32	0·49955	3·82	0·49993
2·33	0·49010	2·83	0·49767	3·33	0·49957	3·83	0·49994
2·34	0·49036	2·84	0·49774	3·34	0·49958	3·84	0·49994
2·35	0·49061	2·85	0·49781	3·35	0·49960	3·85	0·49994
2·36	0·49086	2·86	0·49788	3·36	0·49961	3·86	0·49994
2·37	0·49111	2·87	0·49795	3·37	0·49962	3·87	0·49995
2·38	0·49134	2·88	0·49801	3·38	0·49964	3·88	0·49995
2·39	0·49158	2·89	0·49807	3·39	0·49965	3·89	0·49995
2·40	0·49180	2·90	0·49813	3·40	0·49966	3·90	0·49995
2·41	0·49202	2·91	0·49819	3·41	0·49968	3·91	0·49995
2·42	0·49224	2·92	0·49825	3·42	0·49969	3·92	0·49996
2·43	0·49245	2·93	0·49831	3·43	0·49970	3·93	0·49996
2·44	0·49266	2·94	0·49836	3·44	0·49971	3·94	0·49996
2·45	0·49286	2·95	0·49841	3·45	0·49972	3·95	0·49996
2·46	0·49305	2·96	0·49846	3·46	0·49973	3·96	0·49996
2·47	0·49324	2·97	0·49851	3·47	0·49974	3·97	0·49996
2·48	0·49343	2·98	0·49856	3·48	0·49975	3·98	0·49997
2·49	0·49361	2·99	0·49861	3·49	0·49976	3·99	0·49997
						4·00	0·49997

χ²-DISTRIBUTION

n	α = 0·995	α = 0·99	α = 0·975	α = 0·95	α = 0·05	α = 0·025	α = 0·01	α = 0·005	n
1	0·0000393	0·000157	0·000982	0·00393	3·841	5·024	6·635	7·879	1
2	0·100	0·0201	0·0506	0·103	5·991	7·378	9·210	10·579	2
3	0·00717	0·115	0·216	0·352	7·815	9·348	11·345	12·838	3
4	0·207	0·297	0·484	0·711	9·488	11·143	13·277	14·860	4
5	0·412	0·554	0·831	1·145	11·070	12·832	15·086	16·750	5
6	0·676	0·872	1·237	1·635	12·592	14·449	16·812	18·548	6
7	0·989	1·239	1·690	2·167	14·067	16·013	18·475	20·278	7
8	1·344	1·646	2·180	2·733	15·507	17·535	20·090	21·955	8
9	1·735	2·088	2·700	3·325	16·919	19·023	21·666	23·589	9
10	2·156	2·558	3·247	3·940	18·307	20·483	23·209	25·188	10
11	2·603	3·053	3·186	4·575	19·675	21·920	24·725	26·757	11
12	3·074	3·571	4·404	5·226	21·026	23·337	26·217	28·300	12
13	3·565	4·107	5·009	5·892	22·362	24·736	27·688	29·819	13
14	4·075	4·660	5·629	6·571	23·685	26·119	29·141	31·319	14
15	4·601	5·229	6·262	7·261	24·996	27·488	30·578	32·801	15
16	5·142	5·812	6·908	7·962	26·296	28·845	32·000	34·267	16
17	5·697	6·408	7·564	8·672	27·587	30·191	33·409	35·718	17
18	6·265	7·015	8·231	9·390	28·869	31·526	34·805	37·156	18
19	6·844	7·633	8·907	10·117	30·144	32·852	36·191	38·582	19
20	7·434	8·260	9·591	10·851	31·410	34·170	37·566	39·997	20
21	8·034	8·897	10·283	11·591	32·671	35·479	38·932	41·401	21
22	8·643	9·542	10·982	12·338	33·924	36·781	40·289	42·796	22
23	9·260	10·196	11·689	13·091	35·172	38·076	41·638	44·181	23
24	9·886	10·856	12·401	13·848	36·415	39·364	42·980	45·558	24
25	10·520	11·524	13·120	14·611	37·652	40·646	44·314	46·928	25
26	11·160	12·198	13·844	15·379	38·885	41·923	45·642	48·290	26
27	11·808	12·879	14·573	16·151	40·113	43·194	46·963	49·645	27
28	12·461	13·565	15·308	16·928	41·337	44·461	48·278	50·993	28
29	13·121	14·256	16·047	17·708	42·557	45·722	49·588	52·336	29
30	13·787	14·953	16·791	18·493	43·773	46·979	50·892	53·672	30

FISHER-z VALUES (z_f)

r	0·00	0·01	0·02	0·03	0·04
0·0	0·00000	0·01000	0·02000	0·03001	0·04002
0·1	0·10034	0·11045	0·12058	0·13074	0·14093
0·2	0·20273	0·21317	0·22366	0·23419	0·24477
0·3	0·30952	0·32055	0·33165	0·34283	0·35409
0·4	0·42365	0·43561	0·44769	0·45990	0·47223
0·5	0·54931	0·56273	0·57634	0·59014	0·60415
0·6	0·69315	0·70892	0·72500	0·74142	0·75817
0·7	0·86730	0·88718	0·90764	0·92873	0·95048
0·8	1·09861	1·12703	1·15682	1·18813	1·22117
0·9	1·47222	1·52752	1·58902	1·65839	1·73805

r	0·05	0·06	0·07	0·08	0·09
0·0	0·05004	0·06007	0·07012	0·08017	0·09024
0·1	0·15114	0·16139	0·17167	0·18198	0·19234
0·2	0·25541	0·26611	0·27686	0·28768	0·29857
0·3	0·36544	0·37689	0·38842	0·40006	0·41180
0·4	0·48470	0·49731	0·51007	0·52298	0·53606
0·5	0·61838	0·63283	0·64752	0·66246	0·67767
0·6	0·77530	0·79281	0·81074	0·82911	0·84795
0·7	0·97295	0·99621	1·02033	1·04537	1·07143
0·8	1·25615	1·29334	1·33308	1·37577	1·42192
0·9	1·83178	1·94591	2·09229	2·29756	2·64665

SPEARMAN RANK-CORRELATION COEFFICIENT (r_s)

n	α-Values (one-sided test)	
	0·05	0·01
4	1·000	—
5	0·900	1·000
6	0·829	0·943
7	0·714	0·893
8	0·643	0·833
9	0·600	0·783
10	0·564	0·746
12	0·504	0·701
14	0·456	0·645
16	0·425	0·601
18	0·399	0·564
20	0·377	0·534
22	0·359	0·508
24	0·343	0·485
26	0·329	0·465
28	0·317	0·448
30	0·306	0·432

WILCOXON SIGNED-RANK VALUES

The table gives the values of $T_{1-\alpha}$ for various values of α. The $T_{1-\alpha}$ are such that the probability is approximately (but not more than) α, that $T \leqslant T_{1-\alpha}$. For example, the probability is approximately 0·05 that $T \leqslant 30$ when $n = 15$; and the probability is approximately 0·01 that $T \leqslant 9$ when $n = 12$.

n	$\alpha = 0·05$	$\alpha = 0·025$	$\alpha = 0·01$	$\alpha = 0·005$
4	—	—	—	—
5	0	—	—	—
6	2	0	—	—
7	3	2	0	—
8	5	3	1	0
9	8	5	3	1
10	10	8	5	3
11	13	10	7	5
12	17	13	9	7
13	21	17	12	9
14	25	21	15	12
15	30	25	19	15
16	35	29	23	19
17	41	34	27	23
18	47	40	32	27
19	53	46	37	32
20	60	52	43	37

RANK-SUM CRITICAL VALUES

Numbers in parentheses show sample sizes (n_1, n_2)

	(2, 4)			(4, 4)			(6, 7)	
3	11	0·067	11	25	0·029	28	56	0·26
	(2, 5)		12	24	0·057	30	54	0·051
3	13	0·047		(4, 5)			(6, 8)	
	(2, 6)		12	28	0·032	29	61	0·021
3	15	0·036	13	27	0·056	32	58	0·054
4	14	0·071		(4, 6)			(6, 9)	
	(2, 7)		12	32	0·019	31	65	0·025
3	17	0·028	14	30	0·057	33	63	0·044
4	16	0·056		(4, 7)			(6, 10)	
	(2, 8)		13	35	0·021	33	69	0·028
3	19	0·022	15	33	0·055	35	67	0·047
4	18	0·044		(4, 8)			(7, 7)	
	(2, 9)		14	38	0·024	37	68	0·027
3	21	0·018	16	36	0·055	39	66	0·049
4	20	0·036		(4, 9)			(7, 8)	
	(2, 10)		15	41	0·025	39	73	0·027
4	22	0·030	17	39	0·053	41	71	0·047
5	21	0·061		(4, 10)			(7, 9)	
	(3, 3)		16	44	0·026	41	78	0·027
6	15	0·050	18	42	0·053	43	76	0·045
	(3, 4)			(5, 5)			(7, 10)	
6	18	0·028	18	37	0·028	43	83	0·028
7	17	0·057	19	36	0·048	46	80	0·054
	(3, 5)			(5, 6)			(8, 8)	
6	21	0·018	19	41	0·026	49	87	0·025
7	20	0·036	20	40	0·041	52	84	0·052
	(3, 6)			(5, 7)			(8, 9)	
7	23	0·024	20	45	0·024	51	93	0·023
8	22	0·048	22	43	0·053	54	90	0·046
	(3, 7)			(5, 8)			(8, 10)	
8	25	0·033	21	49	0·023	54	98	0·027
9	24	0·058	23	47	0·047	57	95	0·051
	(3, 8)			(5, 9)			(9, 9)	
8	28	0·024	22	53	0·021	63	108	0·025
9	27	0·042	25	50	0·056	66	105	0·047
	(3, 9)			(5, 10)			(9, 10)	
9	30	0·032	24	56	0·028	66	114	0·027
10	29	0·050	26	54	0·050	69	111	0·047
	(3, 10)			(6, 6)			(10, 10)	
9	33	0·024	26	52	0·021	79	131	0·026
11	31	0·056	28	50	0·047	83	127	0·053

F-DISTRIBUTION $\alpha = 0.05$

ν_2 \ ν_1	1	2	3	4	5	6	7	8	9
1	161·45	199·50	215·71	224·58	230·16	233·99	236·77	238·88	240·54
2	18·513	19·000	19·164	19·247	19·296	19·330	19·353	19·371	19·385
3	10·128	9·5521	9·2766	9·1172	9·0135	8·9406	8·8868	8·8452	8·8123
4	7·7086	6·9443	6·5914	6·3883	6·2560	6·1631	6·0942	6·0410	5·9988
5	6·6079	5·7861	5·4095	5·1922	5·0503	4·9503	4·8759	4·8183	4·7725
6	5·9874	5·1433	4·7571	4·5337	4·3874	4·2839	4·2066	4·1468	4·0990
7	5·5914	4·7374	4·3468	4·1203	3·9715	3·8660	3·7870	3·7257	3·6767
8	5·3177	4·4590	4·0662	3·8378	3·6875	3·5806	3·5005	3·4381	3·3881
9	5·1174	4·2565	3·8626	3·6331	3·4817	3·3738	3·2927	3·2296	3·1789
10	4·9646	4·1028	3·7083	3·4780	3·3258	3·2172	3·1355	3·0717	3·0204
11	4·8443	3·9823	3·5874	3·3567	3·2039	3·0946	3·0123	2·9480	2·8962
12	4·7472	3·8853	3·4903	3·2592	3·1059	2·9961	2·9134	2·8486	2·7964
13	4·6672	3·8056	3·4105	3·1791	3·0254	2·9153	2·8321	2·7669	2·7144
14	4·6001	3·7389	3·3439	3·1122	2·9582	2·8477	2·7642	2·6987	2·6458
15	4·5431	3·6823	3·2874	3·0556	2·9013	2·7905	2·7066	2·6408	2·5876
16	4·4940	3·6337	3·2389	3·0069	2·8524	2·7413	2·6572	2·5911	2·5377
17	4·4513	3·5915	3·1968	2·9647	2·8100	2·6987	2·6143	2·5480	2·4943
18	4·4139	3·5546	3·1599	2·9277	2·7729	2·6613	2·5767	2·5102	2·4563
19	4·3808	3·5219	3·1274	2·8951	2·7401	2·6283	2·5435	2·4768	2·4227
20	4·3513	3·4928	3·0984	2·8661	2·7109	2·5990	2·5140	2·4471	2·3928
21	4·3248	3·4668	3·0725	2·8401	2·6848	2·5757	2·4876	2·4205	2·3661
22	4·3009	3·4434	3·0491	2·8167	2·6613	2·5491	2·4638	2·3965	2·3419
23	4·2793	3·4221	3·0280	2·7955	2·6400	2·5277	2·4422	2·3748	2·3201
24	4·2597	3·4028	3·0088	2·7763	2·6207	2·5082	2·4226	2·3551	2·3002
25	4·2417	3·3852	2·9912	2·7587	2·6030	2·4904	2·4047	2·3371	2·2821
26	4·2252	3·3690	2·9751	2·7426	2·5868	2·4741	2·3883	2·3205	2·2655
27	4·2100	3·3541	2·9604	2·7278	2·5719	2·4591	2·3732	2·3053	2·2501
28	4·1960	3·3404	2·9467	2·7141	2·5581	2·4453	2·3593	2·2913	2·2360
29	4·1830	3·3277	2·9340	2·7014	2·5454	2·4324	2·3463	2·2782	2·2229
30	4·1709	3·3158	2·9223	2·6896	2·5336	2·4205	2·3343	2·2662	2·2107
40	4·0848	3·2317	2·8387	2·6060	2·4495	2·3359	2·2490	2·1802	2·1240
60	4·0012	3·1504	2·7581	2·5252	2·3683	2·2540	2·1665	2·0970	2·0401
120	3·9201	3·0718	2·6802	2·4472	2·2900	2·1750	2·0867	2·0164	1·9588
∞	3·8415	2·9957	2·6049	2·3719	2·2141	2·0986	2·0096	1·9384	1·8799

F-DISTRIBUTION α = 0·05

0	12	15	20	24	30	40	60	120	∞
·88	243·91	245·95	248·01	249·05	250·09	251·14	252·20	253·25	254·32
396	19·413	19·429	19·446	19·454	19·462	19·471	19·479	19·487	19·496
855	8·7446	8·7029	8·6602	8·6385	8·6166	8·5944	8·5720	8·5494	8·5265
644	5·9117	5·8578	5·8025	5·7744	5·7459	5·7170	5·6878	5·6581	5·6281
351	4·6777	4·6188	4·5581	4·5272	4·4957	4·4638	4·4314	4·3984	4·3650
600	3·9999	3·9381	3·8742	3·8415	3·8082	3·7743	3·7398	3·7047	3·6688
365	3·5747	3·5108	3·4445	3·4105	3·3758	3·3404	3·3043	3·2674	3·2298
472	3·2840	3·2184	3·1503	3·1152	3·0794	3·0428	3·0053	2·9669	2·9276
373	3·0729	3·0061	2·9365	2·9005	2·8637	2·8259	2·7872	2·7475	2·7067
782	2·9130	2·8450	2·7740	2·7372	2·6996	2·6609	2·6211	2·5801	2·5379
536	2·7876	2·7186	2·6464	2·6090	2·5705	2·5309	2·4901	2·4480	2·4045
534	2·6866	2·6169	2·5436	2·5055	2·4663	2·4259	2·3842	2·3410	2·2962
710	2·6037	2·5331	2·4589	2·4202	2·3803	2·3392	2·2966	2·2524	2·2064
021	2·5342	2·4630	2·3879	2·3487	2·3082	2·2664	2·2230	2·1778	2·1307
437	2·4753	2·4035	2·3275	2·2878	2·2468	2·2043	2·1601	2·1141	2·0658
935	2·4247	2·3522	2·2756	2·2354	2·1938	2·1507	2·1058	2·0589	2·0096
499	2·3807	2·3077	2·2304	2·1898	2·1477	2·1040	2·0584	2·0107	1·9604
117	2·3421	2·2686	2·1906	2·1497	2·1071	2·0629	2·0166	1·9681	1·9168
779	2·3080	2·2341	2·1555	2·1141	2·0712	2·0264	1·9796	1·9302	1·8780
479	2·2776	2·2033	2·1242	2·0825	2·0391	1·9938	1·9464	1·8963	1·8432
210	2·2504	2·1757	2·0960	2·0540	2·0102	1·9645	1·9165	1·8657	1·8117
967	2·2258	2·1508	2·0707	2·0283	1·9842	1·9380	1·8895	1·8380	1·7831
747	2·2036	2·1282	2·0476	2·0050	1·9605	1·9139	1·8649	1·8128	1·7570
547	2·1834	2·1077	2·0267	1·9838	1·9390	1·8920	1·8424	1·7897	1·7331
365	2·1649	2·0889	2·0075	1·9643	1·9192	1·8718	1·8217	1·7684	1·7110
197	2·1479	2·0716	1·9898	1·9464	1·9010	1·8533	1·8027	1·7488	1·6906
043	2·1323	2·0558	1·9736	1·9299	1·8842	1·8361	1·7851	1·7307	1·6717
900	2·1179	2·0411	1·9586	1·9147	1·8687	1·8203	1·7689	1·7138	1·6541
768	2·1045	2·0275	1·9446	1·9005	1·8543	1·8055	1·7537	1·6981	1·6377
1646	2·0921	2·0148	1·9317	1·8874	1·8409	1·7918	1·7396	1·6835	1·6223
0772	2·0035	1·9245	1·8389	1·7929	1·7444	1·6928	1·6373	1·5766	1·5089
9926	1·9174	1·8364	1·7480	1·7001	1·6491	1·5943	1·5343	1·4673	1·3893
9105	1·8337	1·7505	1·6587	1·6084	1·5543	1·4952	1·4290	1·3519	1·2539
8307	1·7522	1·6664	1·5705	1·5173	1·4591	1·3940	1·3180	1·2214	1·0000

F-DISTRIBUTION $\alpha = 0.01$

$v_2 \backslash v_1$	1	2	3	4	5	6	7	8	
1	4 052·2	4 999·5	5 403·3	5 624·6	5 763·7	5 859·0	5 928·3	5 981·1	6 022·
2	98·503	99·000	99·166	99·249	99·299	99·332	99·356	99·374	99·38
3	34·116	30·817	29·457	28·710	28·237	27·911	27·672	27·489	27·34
4	21·198	18·000	16·694	15·977	15·522	15·207	14·976	14·799	14·65
5	16·258	13·274	12·060	11·392	10·967	10·672	10·456	10·289	10·15
6	13·745	10·925	9·7795	9·1483	8·7459	8·4661	8·2600	8·1016	7·976
7	12·246	9·5466	8·4513	7·8467	7·4604	7·1914	6·9928	6·8401	6·718
8	11·259	8·6491	7·5910	7·0060	6·6318	6·3707	6·1776	6·0289	5·910
9	10·561	8·0215	6·9919	6·4221	6·0569	5·8018	5·6129	5·4671	5·351
10	10·044	7·5594	6·5523	5·9943	5·6363	5·3858	5·2001	5·0567	4·942
11	9·6460	7·2057	6·2167	5·6683	5·3160	5·0692	4·8861	4·7445	4·631
12	9·3302	6·9266	5·9526	5·4119	5·0643	4·8206	4·6395	4·4994	4·387
13	9·0738	6·7010	5·7394	5·2053	4·8616	4·6204	4·4410	4·3021	4·191
14	8·8616	6·5149	5·5639	5·0354	4·6950	4·4558	4·2779	4·1399	4·029
15	8·6831	6·3589	5·4170	4·8932	4·5556	4·3183	4·1415	4·0045	3·894
16	8·5310	6·2262	5·2922	4·7726	4·4374	4·2016	4·0259	3·?8?6	3·780
17	8·3997	6·1121	5·1850	4·6690	4·3359	4·1015	3·9267	3·7910	3·682
18	8·2854	6·0129	5·0919	4·5790	4·2479	4·0146	3·8406	3·7054	3·597
19	8·1850	5·9259	5·0103	4·5003	4·1708	3·9386	3·7653	3·6305	3·522
20	8·0960	5·8489	4·9382	4·4307	4·1027	3·8714	3·6987	3·5644	3·456
21	8·0166	5·7804	4·8740	4·3688	4·0421	3·8117	3·6396	3·5056	3·398
22	7·9454	5·7190	4·8166	4·3134	3·9880	3·7583	3·5867	3·4530	3·345
23	7·8811	5·6637	4·7649	4·2635	3·9392	3·7102	3·5390	3·4057	3·298
24	7·8229	5·6136	4·7181	4·2184	3·8951	3·6667	3·4959	3·3629	3·256
25	7·7698	5·5680	4·6755	4·1774	3·8550	3·6272	3·4568	3·3239	3·217
26	7·7213	5·5263	4·6366	4·1400	3·8183	3·5911	3·4210	3·2884	3·181
27	7·6767	5·4881	4·6009	4·1056	3·7848	3·5580	3·3882	3·2558	3·149
28	7·6356	5·4529	4·5681	4·0740	3·7539	3·5276	3·3581	3·2259	3·119
29	7·5976	5·4205	4·5378	4·0449	3·7254	3·4995	3·3302	3·1982	3·092
30	7·5625	5·3904	4·5097	4·0179	3·6990	3·4735	3·3045	3·1726	3·066
40	7·3141	5·1785	4·3126	3·8283	3·5138	3·2910	3·1238	2·9930	2·887
60	7·0771	4·9774	4·1259	3·6491	3·3389	3·1187	2·9530	2·8233	2·718
120	6·8510	4·7865	3·9493	3·4796	3·1735	2·9559	2·7918	2·6629	2·558
∞	6·6349	4·6052	3·7816	3·3129	3·0173	2·8020	2·6393	2·5113	2·407

F-DISTRIBUTION $\alpha = 0\cdot01$

10	12	15	20	24	30	40	60	120	∞
5 055·8	6 106·3	6 157·3	6 208·7	6 234·6	6 260·7	6 286·8	6 313·0	6 339·4	6 366·0
99·399	99·416	99·432	99·449	99·458	99·466	99·474	99·483	99·491	99·501
27·229	27·052	26·872	26·690	26·598	26·505	26·411	26·316	26·221	26·125
14·546	14·374	14·198	14·020	13·929	13·838	13·745	13·652	13·558	13·463
10·051	9·8883	9·7222	9·5527	9·4665	9·3793	9·2912	9·2020	9·1118	9·0204
7·8741	7·7183	7·5590	7·3958	7·3127	7·2285	7·1432	7·0568	6·9690	6·8801
6·6201	6·4691	6·3143	6·1554	6·0743	5·9921	5·9084	5·8236	5·7372	5·6495
5·8143	5·6668	5·5151	5·3591	5·2793	5·1981	5·1156	5·0316	4·9460	4·8588
5·2565	5·1114	4·9621	4·8080	4·7290	4·6486	4·5667	4·4831	4·3978	4·3105
4·8492	4·7059	4·5582	4·4054	4·3269	4·2469	4·1653	4·0819	3·9965	3·9090
4·5393	4·3974	4·2509	4·0990	4·0209	3·9411	3·8596	3·7761	3·6904	3·6025
4·2961	4·1553	4·0096	3·8584	3·7805	3·7008	3·6192	3·5355	3·4494	3·3608
4·1003	3·9603	3·8154	3·6646	3·5868	3·5070	3·4253	3·3413	3·2548	3·1654
3·9394	3·8001	3·6557	3·5052	3·4274	3·3476	3·2656	3·1813	3·0942	3·0040
3·8049	3·6662	3·5222	3·3719	3·2940	3·2141	3·1319	3·0471	2·9595	2·8684
3·6909	3·5527	3·4089	3·2588	3·1808	3·1007	3·0182	2·9330	2·8447	2·7528
3·5931	3·4552	3·3117	3·1615	3·0835	3·0032	2·9205	2·8348	2·7459	2·6530
3·5082	3·3706	3·2273	3·0771	2·9990	2·9185	2·8354	2·7493	2·6597	2·5660
3·4338	3·2965	3·1533	3·0031	2·9249	2·8422	2·7608	2·6742	2·5839	2·4893
3·3682	3·2311	3·0880	2·9377	2·8594	2·7785	2·6947	2·6077	2·5168	2·4212
3·3098	3·1729	3·0299	2·8796	2·8011	2·7200	2·6359	2·5484	2·4568	2·3603
3·2576	3·1209	2·9780	2·8274	2·7488	2·6675	2·5831	2·4951	2·4029	2·3055
3·2106	3·0740	2·9311	2·7805	2·7017	2·6202	2·5355	2·4471	2·3542	2·2559
3·1681	3·0316	2·8887	2·7380	2·6591	2·5773	2·4923	2·4035	2·3099	2·2107
3·1294	2·9931	2·8502	2·6993	2·6203	2·5383	2·4530	2·3637	2·2695	2·1694
3·0941	2·9579	2·8150	2·6640	2·5848	2·5026	2·4170	2·3273	2·2325	2·1315
3·0618	2·9256	2·7827	2·6316	2·5522	2·4699	2·3840	2·2938	2·1984	2·0965
3·0320	2·8959	2·7530	2·6017	2·5223	2·4397	2·3535	2·2629	2·1670	2·0642
3·0045	2·8685	2·7256	2·5742	2·4946	2·4118	2·3253	2·2344	2·1378	2·0342
2·9791	2·8431	2·7002	2·5487	2·4689	2·3860	2·2992	2·2079	2·1107	2·0062
2·8005	2·6648	2·5216	2·3689	2·2880	2·2034	2·1142	2·0194	1·9172	1·8047
2·6318	2·4961	2·3523	2·1978	2·1154	2·0285	1·9360	1·8363	1·7263	1·6006
2·4721	2·3363	2·1915	2·0346	1·9500	1·8600	1·7628	1·6557	1·5330	1·3805
2·3209	2·1848	2·0385	1·8783	1·7908	1·6964	1·5923	1·4730	1·3246	1·0000

F-DISTRIBUTION $\alpha = 0.025$

ν_1 / ν_2	1	2	3	4	5	6	7	8	9
1	647·79	799·50	864·16	899·58	921·85	937·11	948·22	956·66	963·28
2	38·506	39·000	39·165	39·248	39·298	39·331	39·355	39·373	39·387
3	17·443	16·044	15·439	15·101	14·885	14·735	14·624	14·540	14·473
4	12·218	10·649	9·9792	9·6045	9·3645	9·1973	9·0741	8·9796	8·9047
5	10·007	8·4336	7·7636	7·3879	7·1464	6·9777	6·8531	6·7572	6·6810
6	8·8131	7·2598	6·5988	6·2272	5·9876	5·8197	5·6955	5·5996	5·5234
7	8·0727	6·5415	5·8898	5·5226	5·2852	5·1186	4·9949	4·8994	4·8232
8	7·5709	6·0595	5·4160	5·0526	4·8173	4·6517	4·5286	4·4332	4·3572
9	7·2093	5·7147	5·0781	4·7181	4·4844	4·3197	4·1971	4·1020	4·0260
10	6·9367	5·4564	4·8256	4·4683	4·2361	4·0721	3·9498	3·8549	3·7790
11	6·7241	5·2559	4·6300	4·2751	4·0440	3·8807	3·7586	3·6638	3·5879
12	6·5538	5·0959	4·4742	4·1212	3·8911	3·7283	3·6065	3·5118	3·4358
13	6·4143	4·9653	4·3472	3·9959	3·7667	3·6043	3·4827	3·3880	3·3120
14	6·2979	4·8567	4·2417	3·8919	3·6634	3·5014	3·3799	3·2853	3·2093
15	6·1995	4·7650	4·1528	3·8043	3·5764	3·4147	3·2934	3·1987	3·1227
16	6·1151	4·6867	4·0768	3·7294	3·5021	3·3406	3·2194	3·1248	3·0488
17	6·0420	4·6189	4·0112	3·6648	3·4379	3·2767	3·1556	3·0610	2·9849
18	5·9781	4·5597	3·9539	3·6083	3·3820	3·2209	3·0999	3·0053	2·9291
19	5·9216	4·5075	3·9034	3·5587	3·3327	3·1718	3·0509	2·9563	2·8800
20	5·8715	4·4613	3·8587	3·5147	3·2891	3·1283	3·0074	2·9128	2·8365
21	5·8266	4·4199	3·8188	3·4754	3·2501	3·0895	2·9686	2·8740	2·7977
22	5·7863	4·3828	3·7829	3·4401	3·2151	3·0546	2·9338	2·8392	2·7628
23	5·7498	4·3492	3·7505	3·4083	3·1835	3·0232	2·9024	2·8077	2·7313
24	5·7167	4·3187	3·7211	3·3794	3·1548	2·9946	2·8738	2·7791	2·7027
25	5·6864	4·2909	3·6943	3·3530	3·1287	2·9685	2·8478	2·7531	2·6766
26	5·6586	4·2655	3·6697	3·3289	3·1048	2·9447	2·8240	2·7293	2·6528
27	5·6331	4·2421	3·6472	3·3067	3·0828	2·9228	2·8021	2·7074	2·6309
28	5·6096	4·2205	3·6264	3·2863	3·0625	2·9027	2·7820	2·6872	2·6106
29	5·5878	4·2006	3·6072	3·2674	3·0438	2·8840	2·7633	2·6686	2·5919
30	5·5675	4·1821	3·5894	3·2499	3·0265	2·8667	2·7460	2·6513	2·5746
40	5·4239	4·0510	3·4633	3·1261	2·9037	2·7444	2·6238	2·5289	2·4519
60	5·2857	3·9253	3·3425	3·0077	2·7863	2·6274	2·5068	2·4117	2·3344
120	5·1524	3·8046	3·2270	2·8943	2·6740	2·5154	2·3948	2·2994	2·2217
∞	5·0239	3·6889	3·1161	2·7858	2·5665	2·4082	2·2875	2·1918	2·1136

F-DISTRIBUTION α = 0·025

10	12	15	20	24	30	40	60	120	∞
968·63	976·71	984·87	993·10	997·25	1 001·4	1 005·6	1 009·8	1 014·0	1 018·3
39·398	39·415	39·431	39·448	39·456	39·465	39·473	39·481	39·490	39·498
14·419	14·337	14·253	14·167	14·124	14·081	14·037	13·992	13·947	13·902
8·8439	8·7512	8·6565	8·5599	8·5109	8·4613	8·4111	8·3604	8·3092	8·2573
6·6192	6·5246	6·4277	6·3285	6·2780	6·2269	6·1751	6·1225	6·0693	6·0153
5·4613	5·3662	5·2687	5·1684	5·1172	5·0652	5·0125	4·9589	4·9045	4·8491
4·7611	4·6658	4·5678	4·4667	4·4150	4·3624	4·3089	4·2544	4·1989	4·1423
4·2951	4·1997	4·1012	3·9995	3·9472	3·8940	3·8398	3·7844	3·7279	3·6702
3·9639	3·8682	3·7694	3·6669	3·6142	3·5604	3·5055	3·4493	3·3918	3·3329
3·7168	3·6209	3·5217	3·4186	3·3654	3·3110	3·2554	3·1984	3·1399	3·0798
3·5257	3·4396	3·3299	3·2261	3·1725	3·1176	3·0613	3·0035	2·9441	2·8828
3·3736	3·2773	3·1772	3·0728	3·0187	2·9633	2·9063	2·8478	2·7874	2·7249
3·2497	3·1532	3·0527	2·9477	2·8932	2·8373	2·7797	2·7204	2·6590	2·5955
3·1469	3·0501	2·9493	2·8437	2·7888	2·7324	2·6742	2·6142	2·5519	2·4872
3·0602	2·9633	2·8621	2·7559	2·7006	2·6437	2·5850	2·5242	2·4611	2·3953
2·9862	2·8890	2·7875	2·6808	2·6252	2·5678	2·5085	2·4471	2·3831	2·3163
2·9222	2·8249	2·7230	2·6158	2·5598	2·5021	2·4422	2·3801	2·3153	2·2474
2·8664	2·7689	2·6667	2·5590	2·5027	2·4445	2·3842	2·3214	2·2558	2·1869
2·8173	2·7196	2·6171	2·5089	2·4523	2·3937	2·3329	2·2695	2·2032	2·1333
2·7737	2·6758	2·5731	2·4645	2·4076	2·3486	2·2873	2·2234	2·1562	2·0853
2·7348	2·6368	2·5338	2·4247	2·3675	2·3082	2·2465	2·1819	2·1141	2·0422
2·6998	2·6017	2·4984	2·3890	2·3315	2·2718	2·2097	2·1446	2·0760	2·0032
2·6682	2·5699	2·4665	2·3567	2·2989	2·2389	2·1763	2·1107	2·0415	1·9677
2·6396	2·5412	2·4374	2·3273	2·2693	2·2090	2·1460	2·0799	2·0099	1·9353
2·6135	2·5149	2·4110	2·3005	2·2422	2·1816	2·1183	2·0517	1·9811	1·9055
2·5895	2·4909	2·3867	2·2759	2·2174	2·1565	2·0928	2·0257	1·9545	1·8781
2·5676	2·4688	2·3644	2·2533	2·1946	2·1334	2·0693	2·0018	1·9299	1·8527
2·5473	2·4484	2·3438	2·2324	2·1735	2·1121	2·0477	1·9796	1·9072	1·8291
2·5286	2·4295	2·3248	2·2131	2·1540	2·0923	2·0276	1·9591	1·8861	1·8072
2·5112	2·4120	2·3072	2·1952	2·1359	2·0739	2·0089	1·9400	1·8664	1·7867
2·3882	2·2882	2·1819	2·0677	2·0069	1·9429	1·8752	1·8028	1·7242	1·6371
2·2702	2·1692	2·0613	1·9445	1·8817	1·8152	1·7440	1·6668	1·5810	1·4822
2·1570	2·0548	1·9450	1·8249	1·7597	1·6899	1·6141	1·5299	1·4327	1·3104
2·0483	1·9447	1·8326	1·7085	1·6402	1·5660	1·4835	1·3883	1·2684	1·0000

F-DISTRIBUTION α = 0·005

v_1 / v_2	1	2	3	4	5	6	7	8	9
1	16 211	20 000	21 615	22 500	23 056	23 437	23 715	23 925	24 09
2	198·50	199·00	199·17	199·25	199·30	199·33	199·36	199·37	199·39
3	55·552	49·799	47·467	46·195	45·392	44·838	44·434	44·126	43·882
4	31·333	26·284	24·259	23·155	22·456	21·975	21·622	21·352	21·139
5	22·785	18·314	16·530	15·556	14·940	14·513	14·200	13·961	13·772
6	18·635	14·544	12·917	12·028	11·464	11·073	10·786	10·566	10·39
7	16·236	12·404	10·882	10·050	9·5221	9·1554	8·8854	8·6781	8·5138
8	14·688	11·042	9·5965	8·8051	8·3018	7·9520	7·6942	7·4960	7·3380
9	13·614	10·107	8·7171	7·9559	7·4711	7·1338	6·8849	6·6933	6·541
10	12·826	9·4270	8·0807	7·3428	6·8723	6·5446	6·3025	6·1159	5·9670
11	12·226	8·9122	7·6004	6·8809	6·4217	6·1015	5·8648	5·6821	5·5368
12	11·754	8·5096	7·2258	6·5211	6·0711	5·7570	5·5245	5·3451	5·202
13	11·374	8·1865	6·9257	6·2335	5·7910	5·4819	5·2529	5·0761	4·935
14	11·060	7·9217	6·6803	5·9984	5·5623	5·2574	5·0313	4·8566	4·7173
15	10·798	7·7008	6·4760	5·8029	5·3721	5·0708	4·8473	4·6743	4·5364
16	10·575	7·5138	6·3034	5·6378	5·2117	4·9134	4·6920	4·5207	4·3838
17	10·384	7·3536	6·1556	5·4967	5·0746	4·7789	4·5594	4·3893	4·2535
18	10·218	7·2148	6·0277	5·3746	4·9560	4·6627	4·4448	4·2759	4·1410
19	10·073	7·0935	5·9161	5·2681	4·8526	4·5614	4·3448	4·1770	4·0428
20	9·9439	6·9865	5·8177	5·1743	4·7616	4·4721	4·2569	4·0900	3·9564
21	9·8295	6·8914	5·7304	5·0911	4·6808	4·3931	4·1789	4·0128	3·8799
22	9·7271	6·8064	5·6524	5·0168	4·6088	4·3225	4·1094	3·9440	3·8116
23	9·6348	6·7300	5·5823	4·9500	4·5441	4·2591	4·0469	3·8822	3·7502
24	9·5513	6·6610	5·5190	4·8898	4·4857	4·2019	3·9905	3·8264	3·6949
25	9·4753	6·5982	5·4615	4·8351	4·4327	4·1500	3·9394	3·7758	3·6447
26	9·4059	6·5409	5·4091	4·7852	4·3844	4·1027	3·8928	3·7297	3·5989
27	9·3423	6·4885	5·3611	4·7396	4·3402	4·0594	3·8501	3·6875	3·557
28	9·2838	6·4403	5·3170	4·6977	4·2996	4·0197	3·8110	3·6487	3·5180
29	9·2297	6·3958	5·2764	4·6591	4·2622	3·9830	3·7749	3·6130	3·4832
30	9·1797	6·3547	5·2388	4·6233	4·2276	3·9492	3·7416	3·5801	3·4505
40	8·8278	6·0664	4·9759	4·3738	3·9860	3·7129	3·5088	3·3498	3·2220
60	8·4946	5·7950	4·7290	4·1399	3·7600	3·4918	3·2911	3·1344	3·0083
120	8·1790	5·5393	4·4973	3·9207	3·5482	3·2849	3·0874	3·9330	2·8083
∞	7·8794	5·2983	4·2794	3·7151	3·3499	3·0913	2·8968	2·7444	2·6210

F-DISTRIBUTION $\alpha = 0{\cdot}005$

10	12	15	20	24	30	40	60	120	∞
24 224	24 426	24 630	24 836	24 940	25 044	25 148	25 253	25 359	25 465
199·40	199·42	199·43	199·45	199·46	199·47	199·47	199·48	199·49	199·51
43·686	43·387	43·085	42·778	42·622	42·466	42·308	42·149	41·989	41·829
20·967	20·705	20·438	20·167	20·030	19·892	19·752	19·611	19·468	19·325
13·618	13·384	13·146	12·903	12·780	12·656	12·530	12·402	12·274	12·144
10·250	10·034	9·8140	9·5888	9·4741	9·3583	9·2408	9·1219	9·0015	8·8793
8·3803	8·1764	7·9678	7·7540	7·6450	7·5345	7·4225	7·3088	7·1933	7·0760
7·2107	7·0149	6·8143	6·6082	6·5029	6·3961	6·2875	6·1772	6·0649	5·9505
6·4171	6·2274	6·0325	5·8318	5·7292	5·6248	5·5186	5·4104	5·3001	5·1875
5·8467	5·6613	5·4707	5·2740	5·1732	5·0705	4·9659	4·8592	4·7501	4·6385
5·4182	5·2363	5·0489	4·8552	4·7557	4·6543	4·5508	4·4450	4·3367	4·2256
5·0855	4·9063	4·7214	4·5299	4·4315	4·3309	4·2282	4·1229	4·0149	3·9039
4·8199	4·6429	4·4600	4·2703	4·1726	4·0727	3·9704	3·8655	3·7577	3·6465
4·6034	4·4281	4·2468	4·0585	3·9614	3·8619	3·7600	3·6553	3·5473	3·4359
4·4236	4·2498	4·0698	3·8826	3·7859	3·6867	3·5850	3·4803	3·3722	3·2602
4·2719	4·0994	3·9205	3·7342	3·6378	3·5388	3·4372	3·3324	3·2240	3·1115
4·1423	3·9709	3·7929	3·6073	3·5112	3·4124	3·3107	3·2058	3·0971	2·9839
4·0305	3·8599	3·6827	3·4977	3·4017	3·3030	3·2014	3·0962	2·9871	2·8732
3·9329	3·7631	3·5866	3·4020	3·3062	3·2075	3·1058	3·0004	2·8908	2·7762
3·8470	3·6779	3·5020	3·3178	3·2220	3·1234	3·0215	2·9159	2·8058	2·6904
3·7709	3·6024	3·4270	3·2431	3·1474	3·0488	2·9467	2·8408	2·7302	2·6140
3·7030	3·5350	3·3600	3·1764	3·0807	2·9821	2·8799	2·7736	2·6625	2·5455
3·6420	3·4745	3·2999	3·1165	3·0208	2·9221	2·8198	2·7132	2·6016	2·4837
3·5870	3·4199	3·2456	3·0624	2·9667	2·8679	2·7654	2·6585	2·5463	2·4276
3·5370	3·3704	3·1963	3·0133	2·9176	2·8187	2·7160	2·6088	2·4960	2·3765
3·4916	3·3252	3·1515	2·9685	2·8728	2·7738	2·6709	2·5633	2·4501	2·3297
3·4499	3·2839	3·1104	2·9275	2·8318	2·7327	2·6296	2·5217	2·4078	2·2867
3·4117	3·2460	3·0727	2·8899	2·7941	2·6949	2·5916	2·4834	2·3689	2·2469
3·3765	3·2111	3·0379	2·8551	2·7594	2·6601	2·5565	2·4479	2·3330	2·2102
3·3440	3·1787	3·0057	2·8230	2·7272	2·6278	2·5241	2·4151	2·2997	2·1760
3·1167	2·9531	2·7811	2·5984	2·5020	2·4015	2·2958	2·1838	2·0635	1·9318
2·9042	2·7419	2·5705	2·3872	2·2898	2·1874	2·0789	1·9622	1·8341	1·6885
2·7052	2·5439	2·3727	2·1881	2·0890	1·9839	1·8709	1·7469	1·6055	1·4311
2·5188	2·3583	2·1868	1·9998	1·8983	1·7891	1·6691	1·5325	1·3637	1·0000

ANSWERS TO EXERCISES

Chapter II

1. (a) 42·45–44·45, 44·45–46·45, 46·45–48·45; (b) 43·45, 45·45, 47·45
2. (a) 0·4705–0·4755, 0·4755–0·4805, 0·4805–0·4855; (b) 0·4730, 0·4780, 0·4830
3. (a) 1·39–1·42, 1·43–1·46, 1·47–1·50; (b) 1·405, 1·445, 1·485
4. (a) 110·5–119·5, 119·5–128·5, 128·5–137·5; (b) 111–119, 120–128, 129–137
5. (a) 2·455–2·495, 2·495–2·535, 2·535–2·575; (b) 2·46–2·49, 2·50–2·53, 2·54–2·57
6. (a) 39·85, 44·35, 48·85; (b) 37·6–42·0, 42·1–46·5, 46·6–51·0; (c) 37·60–42·09, 42·10–46·59, 46·60–51·09
7. There is no unique answer.
8. The class boundaries are 199·5–219·5, 219·5–239·5, 239·5–259·5, 259·5–279·5, 279·5–299·5, 299·5–319·5, 319·5–339·5
   The class limits are 200–219, 220–239, 240–259, 260–279, 280–299, 300–319, 320–339
   The class marks are 209·5, 229·5, 249·5, 269·5, 289·5, 309·5, 329·5
9. The class boundaries are 4·85–5·95, 5·95–7·05, 7·05–8·15, 8·15–9·25, 9·25–10·35, 10·35–11·45, 11·45–12·55, 12·55–13·65
   The class limits are 4·9–5·9, 6·0–7·0, 7·1–8·1, 8·2–9·2, 9·3–10·3, 10·4–11·4, 11·5–12·5, 12·6–13·6
   The class marks are 5·40, 6·50, 7·60, 8·70, 9·80, 10·90, 12·00, 13·10
10. There is no unique answer.

Chapter III

1. (a) 9   (b) 50·5   (c) 1·586   (d) 125·8
2. (a) 8   (b) 9, 10   (c) no mode   (d) 7, 8
3. (a) 10   (b) 47   (c) 1·603   (d) 134·8
4. 38
5. 35
6. (a) 9   (b) 47·8   (c) 1·599   (d) 130·0
7. 368·8
8. The values of both the median and the mean depend upon how the data are classified. The median of the *unclassified* data is 46·6; the mean of the *unclassified* data is 46·8. The mean and median of the classified data should be close to these values for any choice of class interval and class boundaries.
9. Same comment as in Exercise 8 above. The mean of the *unclassified* data is 1·007.

Chapter IV

1. The range, mean absolute deviation, variance, and standard deviation are given in each part in that order
   (a) 8, 2·29, 8, 2·83
   (b) 39, 11·17, 199·0, 14·11
   (c) 0·173, 0·048, 0·0039, 0·0624
   (d) 48·7, 13·75, 340·5, 18·5
   (e) 6, 1·35, 3·81, 1·95
   (f) 7, 1·6, 4·81, 2·19
2. 208·6
3. The variance of the *unclassified* data is 53·21.
4. The variance of the *unclassified* data is 0·000265.
   The standard deviation of the unclassified data is 0·0163.

Chapter V

1. (a) $\frac{1}{13}$ (b) $\frac{1}{13}$ (c) $\frac{3}{13}$ (d) $\frac{1}{2}$ (e) $\frac{1}{4}$
2. (a) $\frac{1}{15}$ (b) $\frac{7}{15}$ (c) $\frac{8}{15}$ (d) $\frac{1}{3}$
3. (a) $\frac{1}{9}$ (b) $\frac{5}{18}$ (c) $\frac{7}{12}$
4. (a) $\frac{8}{15}$ (b) 0 mutually exclusive   (c) $\frac{7}{15}$
5. (a) 0·15   (b) 0·65   (c) 0·35
6. (a) $\frac{25}{102}$ (b) $\frac{25}{102}$ (c) $\frac{25}{51}$ (d) $\frac{26}{51}$
7. $\frac{38}{85}$
8. $\frac{14}{5525}$
9. $\frac{1}{4}$
10. $\frac{1}{108}$
11. (a) 120   (b) 120   (c) 120   (d) 42   (e) 6   (f) 56
12. (a) 60   (b) 120   (c) 120
13. 3 603 600
14. 12 650
15. $_{52}C_5$
16. (a) $\frac{1}{6}$ (b) $\frac{1}{2}$
17. (a) 0·082   (b) 0·329   (c) 0·273
18. (a) 0·211   (b) 0·262   (c) 0·375   (d) 0·688
19. (a) $\frac{5}{16}$ (b) $\frac{21}{32}$
20. 1·5

Chapter VI

1. (a) 0·45994   (b) 0·03586   (c) 0·49010   (d) 0·39065   (e) 0·27935
2. (a) 0·25175   (b) 0·33891   (c) 0·39435   (d) 0·49506   (e) 0·42647
3. (a) 0·95994   (b) 0·53586   (c) 0·99010   (d) 0·89065   (e) 0·77935
4. (a) 0·24825   (b) 0·16109   (c) 0·10565   (d) 0·00494   (e) 0·07353
5. (a) 0·34974   (b) 0·06852   (c) 0·51561   (d) 0·13591   (e) 0·46029
   (f) 0·81648   (g) 0·12507   (h) 0·69146   (i) 0·10935
6. (a) 0·84134   (b) 0·80785   (c) 0·57142   (d) 0·69146   (e) 0·62845
   (f) 0·56749
7. (a) 0·27701   (b) 0·12924   (c) 0·84375   (d) 0·35946
8. (a) 0·91149   (b) 0·99492
9. (a) 0·97128   (b) 0·75804   (c) 0·13567   (d) 0·19803   (e) 0·23582
10. 0·01923
11. 0·00097, $P_N(x \leqslant 34·5)$
12. 0·38292, $P_N(48 \leqslant x \leqslant 52)$

Chapter VII

1. $z = 1·67$; therefore, reject H_0.
2. $z = 2·60$; yes, conclude that the coin is biased.
3. $z = -1·80$; reject H_0 at the 0·05 level.
4. $z = -1·64$; accept H_0 at the 0·05 level.
5. Type I error is about 0·005. The Type II error, when $\pi = 0·5$, is approximately 0·69.
6. $z = -1·25$; accept H_0.
7. $z = 1·29$; accept H_0 at the 0·05 level.

Chapter VIII

1. $t = 2·63$; reject H_0.
2. $\bar{x} = 11·25$, $s^2 = 4·25$; we find $t = 2·38$; accept H_0.
3. $z = -8$; reject H_0.
4. $z = 2·70$; reject H_0 at the 0·01 level.

242 *Statistics Made Simple*

5. $z = 0.95$; accept H_o; $\pi_1 - \pi_2 \leq 0.10$.
6. $z = 2.45$; reject H_o: $\pi_1 = \pi_2$ at the 0.01 level, and conclude that there is a difference between the proportions.
7. $z = 0.534$; accept H_o.
8. $s_p^2 = 43.33$; $t = 1.20$; accept H_o.

Chapter IX

1. $r = 0.685$.
2. $r = -0.744$.
3. $z = -0.74$; accept H_o.
4. $z = 0.99$; accept H_o.
5. $z = -0.70$; accept H_o, when $\alpha = 0.01$.
6. $\bar{y} = 19.625$; $b = 0.814$; $\hat{y}_i = -2.35 + 0.814x_i$.
7. $s^2 = 27.38$.
8. $t = 2.31$; accept H_o when $\alpha = 0.025$.
9. $t = -0.53$; accept H_o, when $\alpha = 0.10$.
10. $t = 0.88$; accept H_o at 0.10 level.
11. $\bar{y} = 11.333$; $b = -1.063$; $\hat{y}_i = 27.63 - 1.063x_i$.
12. 17.00, 15.94, 11.69, 9.56, 7.43, 6.37.
13. $\sum(y_i - \hat{y}_i)^2 = 12.995$; $s = 1.802$.
14. $t = -5.45$.

Chapter X

1. $(21.71, 28.29)$, $(21.08, 28.92)$, $(19.84, 30.16)$
2. $(182.2, 197.8)$
3. $(12.21, 19.39)$
4. $(0.593, 0.807)$
5. $(0.524, 0.576)$, $(0.519, 0.581)$, $(0.509, 0.591)$
6. $(-5.15, 13.15)$
7. $(-13.24, 2.24)$
8. $(53.63, 70.37)$
9. $(-0.013, 0.213)$
10. $(-0.171, 0.271)$
11. $(-0.020, 0.120)$
12. $(-0.46, 0.84)$
13. $(-0.88, -0.51)$
14. $(0.04, 0.87)$, $(0.46, 0.71)$, $(0.53, 0.66)$

Chapter XI

1. No unique answer.
2. $\chi^2 = 10.17$; reject H_o at the 0.05 level. The theory is not compatible with the observed values.
3. $\chi^2 = 19.10$; conclude that the criteria are not independent.
4. $\chi^2 = 344.66$; conclude that the educational level is not independent of the number of books read.
5. $z = -1.02$; accept H_o.
6. $z = -0.75$; accept H_o at the 0.05 level.
7. The sum of the positive ranks is 109; accept H_o at the 0.05 level.
8. The sum of the positive ranks is 56.5; accept H_o at the 0.05 level.
9. $R = 64.5$; reject H_o at the 0.05 level.
10. $R = 60$; accept H_o at the 0.05 level.

Chapter XII

1.

Source	d.f.	S.S.	M.S.
Among	2	40	20
Within	12	60	5
Total	14	100	—

$F = 4.00$; reject H_o at the 0·05 level.

2.

Source	d.f.	S.S.	M.S.
Among	3	78·62	26·21
Within	16	95·18	5·95
Total	19	173·80	—

$F = 4.41$; reject H_o at the 0·05 level.

3.

Source	d.f.	S.S.	M.S.
Rows	3	202·19	67·40
Cols.	3	134·19	44·73
Within	9	302·56	33·62
Total	15	638·94	—

$F = 2.00$; accept that row means are equal.
$F = 1.33$; accept that column means are equal.

4.

Source	d.f.	S.S.	M.S.
Rows	4	52·92	13·23
Cols.	2	38·80	19·40
Within	8	173·88	21·74
Total	14	265·60	—

$F = 0.61$; accept that row means are equal.
$F = 0.89$; accept that column means are equal, at the 0·01 level.

Index